Citrus Press
Titusville, FL
407-264-2200
Jim Burnett

ERRATA

Page 10, section 1.05 DISCRETE DATA, 4th line from bottom:

"representing continuous functions. If, on the other hand, we are careless, the relationship of the discrete function to what we suppose to be the underlying continuous curve may be completely unfounded (perhaps we can discuss such things in another book).

Page 31, COMMENTARY, last paragraph:

"   In addition to these comments, the DFT deals with arrays of discrete, digital data.   There are no smooth, continuous curves in a computer."

Page 106, section 6.4 OMITTING THE NEGATIVE FREQUENCIES, 1st para.:

"   As we noted back in chapter 3,when the input is limited to real numbers, we do not need to compute the negative frequencies—they are only the complex conjugates of the real frequencies."

Page 130, last paragraph, first sentence.:

"   Based on our benchmark of 1μsec per operation, it would take approximately 0.022 seconds to transform the 2048 data point array we discussed in the "audio" example at the end of chapter 6—that's 0.044 seconds for transform *and* reconstruction."

Page 151, section 8.5 THE INVERSE TRANSFORM 1st para.:

"   We already have the inverse FFT of course—it's the same algorithm we just developed (see chapter 4).   We need only make forward/inverse *scale factor and sign* changes, and minor changes in program control."

Page 152, 2nd para., 6th line:

"   If we replace the "2" in these equations with a variable "SK1", we may set SK1 to 2 when doing the forward transform and to 1 when doing the inverse. For the sign change we simply introduce the variable K6:

```
187 C(T0,J1+I)=(C(T1,I+J)+C(T1,K)*KC(KT)-K6*S(T1,K)*KS(KT))/SK1
188 S(T0,J1+I)=(S(T1,I+J)+K6*C(T1,K)*KS(KT)+S(T1,K)*KC(KT))/SK1

194 C(T0,J1+I)=(C(T1,I+J)+C(T1,K)*KC(KT)-K6*S(T1,K)*KS(KT))/SK1
195 S(T0,J1+I)=(S(T1,I+J)+K6*C(T1,K)*KS(KT)+S(T1,K)*KC(KT))/SK1
```

at line 612 we set SK1 = 2 and K6 to positive 1 when we set the T0/T1 flags:

```
612 T0=1:T1=0:SK1=2:K6=1
```

Page 153, top of page:

"   Then, at line 700, we write the inverse transform routine.

```
700 REM *** INVERSE TRANSFORM ***
710 SK1 = 1:K6 = -1
712 CLS:PRINT "TIME    AMPLITUDE    NOT USED         ";
714 PRINT "TIME    AMPLITUDE    NOT USED":PRINT:PRINT
720 GOSUB 106
730 RETURN
```

We set SK1 to 1 (line 710) thereby removing the forward transform scale factor, K6 to -1, and print a new heading for the output data."

Understanding
the

FFT

A Tutorial on
the Algorithm &
Software for Laymen,
Students, Technicians
& Working Engineers

Anders E. Zonst

Citrus Press
Titusville, Florida

Publishers Cataloging-in-Publication Data

Zonst, Anders E.
      Understanding the FFT—A Tutorial on the Algorithm & Software
      for Laymen, Students, Technicians and Working Engineers/ A.E. Zonst
      p. cm.
      Includes bibliographical references and index
      1. Fourier analysis. 2. Fourier transformations
      3. Fourier Transformations—data processing I. Title

QA403.Z658   1995   515.7'23   95-67942
ISBN 0-9645681-8-7

Library of Congress Catalog Card Number: 95-67942

International Standard Book Number 0-9645681-8-7

Printed in the United States of America 10 9 8 7 6 5 4 3 2 1

To Shirley

ACKNOWLEDGEMENTS

A special thanks to Renée, Heather and Maureen. Without their help you would have had to suffer countless instances of bad grammar, awkward syntax and incorrect spelling. Of much more importance, from my point of view, they have nice personalities and are fun to be around.

CONTENTS

Part II - The FFT

PROLOGUE

"Considering how many fools can calculate, it is surprising that it should be thought either a difficult or a tedious task for any other fool to learn how to master the same tricks.

Some calculus-tricks are quite easy. Some are enormously difficult. The fools who write the text books of advanced mathematics—and they are mostly clever fools—seldom take the trouble to show you how easy the easy calculations are. On the contrary, they seem to desire to impress you with their tremendous cleverness by going about it in the most difficult way.

Being myself a remarkably stupid fellow, I have had to unteach myself the difficulties, and now beg to present to my fellow fools the parts that are not hard. Master these thoroughly, and the rest will follow. What one fool can do, another can." (Prologue to *Calculus Made Easy* by Silvanus P. Thompson, F.R.S., 1910)

Even though a great many years had passed since I first obtained a copy of Thompson's magical little book (twenty-eighth printing of the second edition, Macmillan Publishing Company, 1959), I nonetheless recognized this prologue when a part of it appeared recently on the front page of a book. The reader should understand that Professor Thompson wasn't simply being sarcastic. His intention was, beyond question, to throw a lifeline to floundering students. His goal was to provide an introduction to that powerful tool known as The Calculus; to provide a bridge for those who had been victimized by their teachers and texts. Lest anyone mistake his true feelings, he adds the following in the epilogue: "...One other thing will the professed mathematicians say about this thoroughly bad and vicious book: that the reason why it is *so easy* is because the author has left out all the things that are really difficult. And the ghastly fact about this accusation is that—*it is true!* That is, indeed, why the book has been written—written for the legion of innocents who have hitherto been deterred from acquiring the elements of the calculus by the stupid way in which its teaching is almost always presented. Any subject can be made repulsive by presenting it bristling with difficulties. The aim of this book is to enable beginners to learn its language, to acquire familiarity with its endearing simplicities, and to grasp its powerful methods of problem solving, without being compelled to toil through the intricate out-of-the-way (and mostly irrelevant) mathematical gymnastics so dear to the unpractical mathematician..." (From the Epilogue and Apology of *Calculus Made Easy* by Silvanus P. Thompson, 1910. Apparently some things never change.)

I cannot be sure that the coincidence of Thompson's prologue, printed boldly on the front page of an exemplary treatise on Fourier Analysis, was the sole motivation for this book—I had already considered just such an essay. Still, if Thompson's ghost had appeared and spoken to me directly, my task would not have been clearer. Allow me to explain: This book is intended to help those who would like to understand the Fast Fourier Transform (FFT), but find the available literature too burdensome. It is born of my own frustration with the papers and texts available on the FFT, and the perplexing way in which this subject is usually presented. Only after an unnecessarily long struggle did I find that the FFT was actually simple—*incredibly simple*. You do not need to understand advanced calculus to understand the FFT—you certainly do not need deliberately obscure notation and symbols that might be more appropriate to the study of archeology. The simple truth is that the FFT could easily be understood by any high school student with a grasp of trigonometry. Understand, then, that I hold heart-felt sympathy with Thompson's iconoclasm. In fact, if you swap "FFT" for "Calculus," Thompson's strong words express my own feelings better than I am capable of expressing them myself.

But there is another, perhaps better, reason for this book. Today, systems using the FFT abound—real systems—solving real problems. The programmers, engineers and technicians who develop, use, and maintain these systems need to *understand* the FFT. Many of these people have long since been "excommunicated" from the specialized groups who discuss and write about this subject. It may be acceptable for professional scholars to communicate via abstruse hieroglyphics, but working engineers and technicians need a more direct route to their tools. This book aims to provide a direct route to the FFT.

INTRODUCTION

This book is written in two parts—an introduction to (or review of) the DFT, and an exposition of the FFT. It is a little book that can be read in a few evenings at most. Recognizing this, I recommend that you start from the beginning and read it all— each chapter builds on all that has preceded. If you are already familiar with the DFT the first four chapters should read comfortably in a single evening.

I have gone as far as I can to make this subject accessible to the widest possible audience, including an appendix 1.1 which provides a "refresher" on the BASIC language. After that, the programs in Part I start out very simply with detailed explanations of each line of code in the text.

My reason for including these features is that, some years ago (before the advent of the personal computer), there was a period of several years in my life when I was "computer-less." When I once again obtained access to a computer I was shocked to find that I had forgotten the commands and rules for programming (even in BASIC). To my great relief a few hours at a keyboard (with a BASIC programming manual in hand) brought back enough to get me up and running. Appendix 1.1, and the programs of part 1, are designed to accomplish the same thing with much less pain.

In addition to these comments, I should point out that the programs presented in this book are intended to be typed into a computer and run—they actually work. If you don't like to type, a disk with all the program listings can be furnished for $5.00 (which includes postage and handling).

Very well then, the first topic we will consider is: "What, actually, is the Digital Fourier Transform?"

CHAPTER I
STARTING AT THE BOTTOM

It has been said that a good definition first throws the thing to be defined into a very large pool (i.e. a very broad category) and then pulls it out again (i.e. describes the unique characteristics that differentiate it from the other members of that category). That is the approach we will use in tackling the question; "What, exactly, is the Fourier series?"

1.01 APPROXIMATION BY SERIES

When we first encounter mathematical functions they are defined in simple, direct terms. The common trigonometric functions, for example, are defined with respect to a right triangle:

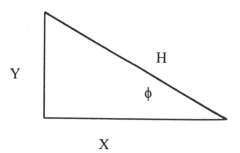

$$Sin(\emptyset) = Y/H \text{ --------------------------------} \quad (1.1)$$

$$\emptyset = \text{angle } \emptyset$$
$$Y = \text{height}$$
$$H = \text{hypotenuse}$$

$$Cos(\emptyset) = X/H \text{ --------------------------------} \quad (1.2)$$

$$X = \text{base}$$

$$Tan(\emptyset) = Y/X \text{ --------------------------------} \quad (1.3)$$

Shortly thereafter we learn that these functions may also be expressed as a series of terms:

$$Sin(x) = x - x^3/3! + x^5/5! - x^7/7! +... \quad (1.4)$$

$$x = \text{angle in radians}$$
$$3!, 5!, \text{ etc.} = \text{3 factorial, 5 factorial, etc.}$$

$$Cos(x) = 1 - x^2/2! + x^4/4! - x^6/6! +... \quad (1.5)$$

These *power series* are known as Maclauren/Taylor series and may be derived for all the commonly used trigonometric and transcendental functions.

1.02 THE FOURIER SERIES

The Fourier series is a *trigonometric series*. Specifically, it is a series of sinusoids (plus a constant term), whose amplitudes may be determined by a certain process (to be described in the following chapters). Equation (1.6) states the Fourier series explicitly; unfortunately, this compact notation cannot reveal the

$$F(x) = A_0 + A_1 Cos(x) + A_2 Cos(2x) + A_3 Cos(3x) + ...$$
$$+ B_1 Sin(x) + B_2 Sin(2x) + B_3 Sin(3x) + ... \qquad (1.6)$$

incredible mathematical subtlety contained within. The Fourier series, like the Taylor/Maclauren series shown earlier, approximates functions, but it has a different derivation and a different purpose. Rather than being a means of evaluating sines, cosines, etc., at a single point, it serves as a "transformation" for the whole of a given, arbitrary, function.

This, then, is the general pool that we have thrown our Fourier Transform into, but we are at risk here of making the pool so obscure it will require more definition than our definition itself. The newcomer may well ask; "What is this *transformation* you speak of?" Apparently we are going to transform the original function into another, different, function—but what is the new function and why do we bother? Does the transformed function have some special mathematical properties? Can we still obtain the same information provided by the original function? The answer is yes to both of these questions but we will come to all of that later; for now we may say that the transform we are referring to, in its digital form, provides a mathematical tool of such power and scope that it can hardly be exceeded by any other development of applied mathematics in the twentieth century.

Now we move to the second part of our definition—we must pull the defined topic out of the pond again. This part of the definition requires that we speak carefully and use our terms precisely, for now we hope to reveal the specific nature of the Fourier Transform. We will begin with a couple of definitions that will be used throughout the remainder of this book.

1.03 FUNCTIONS

The term *function* (or *single valued function*) is, in a sense, a "loose" term (i.e. it describes a very simple notion that can be fulfilled many different ways). It only implies a set of *ordered pairs* of numbers (x,y), where the second number (y, the dependent variable) is a *unique* value which corresponds to the first number (x, the independent variable). Now, obviously, the equations encountered in engineering and physics can provide sets of numbers which fulfill this definition, *but so will any simple list of numbers*. It is *not* necessary to know the equation that relates the dependent to the independent variable, nor even that the two numbers be related by an equation at all! This "looseness" is essential if the term "function" is to cover the practical work that is done with a Digital Fourier Transform (DFT), for it is seldom that the information obtained by digitizing a signal can be described by an equation.

1.03.1 Discontinuous Functions

There are functions, encountered frequently in technical work (and especially in Fourier Analysis), that are difficult to describe in a few words. For example, the "Unit Square Wave", must be defined in some manner such as the following:

$$f(x) = 1 \text{ [for } 0 \leq x < x_1] \text{ ------------------} \qquad (1.7)$$

and:

$$f(x) = -1 \text{ [for } x_1 \leq x < 2x_1] \text{ -----------------} \qquad (1.8)$$

We make no statement about this function outside the interval of $0 \leq x < 2x_1$. This interval is referred to as the "domain of defini-tion" or simply the "domain" of the function. We will have more to say about the domain of a function and its transform shortly, but for now let's continue to investigate discontinuous functions. We require two separate equations to describe this Square Wave function, but we also need some explanation: At the point $x = x_1$ the first equation ends and the second equation begins-there is no "connection" between them. The function is discontinuous at the point where it jumps from +1 to -1. It is sometimes suggested that these two equations be connected by a straight, vertical line of "infinite slope", but this "connection" cannot be allowed. A con-necting line of infinite slope would have more than one value (in fact, an infinite number of values) at the "point" of transition. The definition of a *single valued function* requires a single "unique" value of the dependent variable for any value of the independent variable.

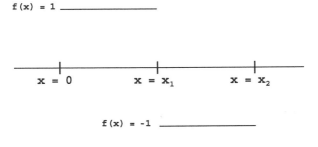

f(x) = Unit Square Wave

Mathematically, functions such as the unit square wave must remain discontinuous; but, physically, such discontinuities are not realizable. All voltage square waves measured in a laboratory, for example, will have *finite* rise and fall times.

1.04 THE FOURIER SERIES

Our original question was; "What, exactly, is a Fourier series?" We stated back on page 3 that it is a series of sinusoids, and as an aid to the intuition, it is frequently shown that a square wave may be approximated with a series of sine waves. Now, these sinusoids are, in general, referred to as the "harmonics" of the wave shape, except that a sine wave which just fits into the domain of definition (i.e. one cycle fits *exactly* into the waveform domain) is called the *fundamental* (see fig. 1.1A). If two cycles fit into this interval they are called the second harmonic (note that there is no first harmonic—that would correspond to the fundamental). If three cycles of sine wave fit into the interval they are called the third harmonic, etc., etc. A square wave, as described in the preceding section, consists of, in addition to the fundamental, only the "odd" numbered harmonics (i.e. 3rd, 5th, 7th, etc., etc.), all of whose amplitudes are inversely proportional to their harmonic number. Caveat: To represent a square wave perfectly by Fourier series, an infinite number of harmonic components would be required. That is to say, the Fourier series can never perfectly reproduce such functions; however, it *can* reproduce them to any desired degree of accuracy, just as the Taylor series shown in eqns.

(1.4) and (1.5) will converge to any desired accuracy (another caveat: convergence of the Fourier series is not a simple subject—but that discussion diverges from our immediate purpose).

In figure 1.1 we show how a summation of odd harmonics begins to form a square wave. Even though we only sum in the first four components, it is apparent that a square wave is beginning to form.

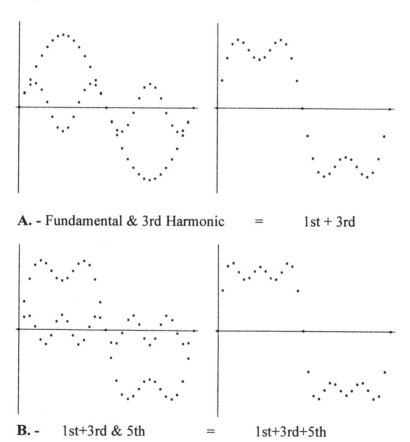

A. - Fundamental & 3rd Harmonic   =   1st + 3rd

B. -   1st+3rd & 5th   =   1st+3rd+5th

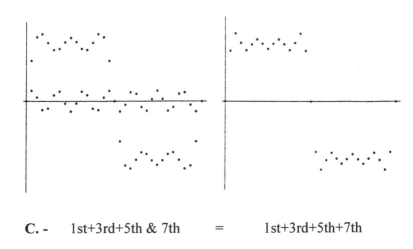

C. -   1st+3rd+5th & 7th   =   1st+3rd+5th+7th

Figure 1.1 - Construction of a Square Wave.

In figure 1.2 below, we show the results of summing in 11 components and 101 components. We note that with 101 components the approximation is very good although not perfect. The basic idea illustrated here, however, of approximating the wave-

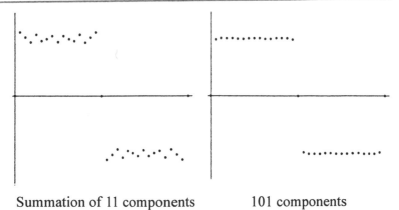

Summation of 11 components   101 components

Figure 1.2

form of a function by summing harmonically related sinusoids, is the fundamental idea underlying the Fourier series. The implication is that *any* function may be approximated by summing harmonic components (is this true?)

1.05 DISCRETE DATA

Let's take the time now to point out a distinct characteristic of the above "curves." In this book the graphs will usually show what actually happens in digital systems—they will display mathematical "points" plotted at regular intervals. While the Unit Square Wave of section 1.03.1 is discontinuous at the transition point, the functions of figs. 1.1 and 1.2 are discontinuous at every point. This is not a trivial phenomenon—a series of discrete data points is not the same thing as a continuous curve. We suppose the "continuous curves," from which we extract discrete data points, are still somehow represented; but, this supposition may not be justified. This characteristic of sampled data systems, and of digital systems in general, creates idiosyncracies in the DFT that are not present in the continuous Fourier series (e.g. the subtleties of convergence, which were hinted at above, are obviated by the *finite series* of the DFT). Put simply, our situation is this: If we treat discrete functions carefully, we may think of them as representing underlying linear functions. If, on the other hand, we are careless, the relationship of the discrete function to what we suppose to be the underlying linear function may be completely unfounded (perhaps we can discuss such things in another book).

1.06.1 COMPUTER PROGRAMS

It is anticipated that most readers will have some familiarity with computer programming; but, if not, don't be intimidated. We will start with very simple examples and explain everything we are doing. Generic BASIC is pretty simple, and the examples will gradually increase in difficulty so that you should have no trouble following what we are doing. Understand, however, that these programs are not just exercises—they *are* the book in the same way that this text *is* the book. This is what you are trying to learn. Type these programs into a computer and run them—experiment with them—(but be careful, this stuff can be addictive). If you have no familiarity with BASIC at all, or if you have not used BASIC in years, you might want to read Appendix 1.1 at this time.

1.06.2 PROGRAM DESCRIPTION

The "square wave" illustration of Figures 1.1 and 1.2 is our first programming example. DFT 1.0 (next page) is essentially the routine used to generate those figures, and its operation is completely illustrated by those figures.

BASIC ignores remarks following REM statements (see line 10). Line 12 asks how many terms we want to sum together and assigns this value to N. Line 20 defines the value of PI. In line 30 we set up a loop that steps the independent variable through

```
10 REM    *** DFT 1.0 - GENERATE SQUARE WAVE  ***
12 INPUT "NUMBER OF TERMS";N
20 PI = 3.14159265358#
30 FOR I = 0 TO 2*PI STEP PI/8
32 Y=0
40 FOR J=1 TO N STEP 2: Y=Y+SIN(J*I)/J: NEXT J
50 PRINT Y
60 NEXT I
70 END
```

Fig. 1.3 - DFT 1.0

2*PI radians (i.e. a full cycle of the fundamental) in increments of PI/8 (if you do not understand the loop structure set up between lines 30 and 60 read appendix 1.1 now). The "loop counter" for this loop is the variable I, which we also use as the independent variable for the equation in line 40. The loop counter I steps in increments of PI/8, yielding 16 data points. Line 32 clears the variable Y which will be used to "accumulate" (i.e. sum together) the values calculated in the "one line loop" at line 40. Mathematically, line 40 solves the following equation:

$$Y = SIN(I*J)/J \textbf{ (for all odd J)} \quad ------------ \qquad (1.9)$$

$$J = \text{harmonic number}$$
$$I = \text{argument of the fundamental}$$

Note that division by the harmonic number (J) yields values inversely proportional to the harmonic number. Line 40 is the heart of the program. It is a loop which counts the variable J "up" from 1 to the number of harmonic terms we requested at line 12 (i.e. "N"). It should be apparent that we are computing the contribution of each harmonic to the waveshape at a given point on the x axis (refer to fig. 1.1 if this is not clear). Each time we pass through the loop, J is incremented by *two*, so that it takes on only odd harmonic values. Each time through the loop we will sum the following into the variable Y:

1) The value it already has (which is zero the first time through the loop), plus...

2) SIN(I*J)/J.

Since I is the value of the argument (in radians) of the fundamental, it will be apparent that I*J represents the "distance" we have progressed through the Jth harmonic component.

When all of the harmonic terms have been summed in (i.e. J = N), we move down to line 50 and print the result. At line 60 we encounter the NEXT I statement, jump back to line 30, increase the variable I by PI/8 radians, and compute all of the harmonic terms for the next position along the x axis.

1.07 EXPERIMENTATION/PRACTICE

The reader should type the above program into a computer and run it. Once you have it working, try variations—sum up hundreds (or even thousands) of harmonic components—modify the mathematical function itself. A simple modification will produce a "ramp" or "sawtooth" function (as opposed to the squarewave). Simply allow the loop counter in line 40 (i.e. J) to step through *all* of the harmonic numbers (i.e. remove the optional "STEP 2" statement in line 40).

Figures 1.4.x (where "x" indicates "don't care") show some of the waveshapes that may be obtained along with the variations required to the equation in line 40. These curves illustrate the effect obtained by simple modifications of the "spectrum" (i.e. the amplitudes and phases of the harmonic components). After playing with this program, and generating a sufficiently large number of functions, we might suspect that any of the common waveshapes encounter in engineering could be produced by selecting the correct spectrum. There are an infinite number of

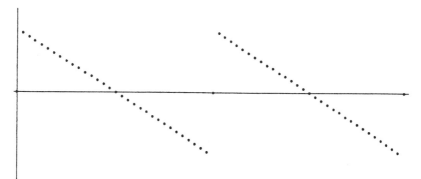

Fig. 1.4.1 - Y=Y+Sin(J*I)/J for all terms.

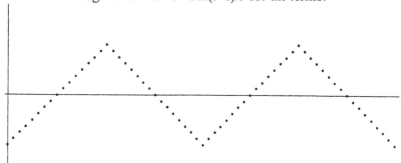

Fig. 1.4.2 - Y=Y-Cos(J*I)/(J*J) for odd terms.

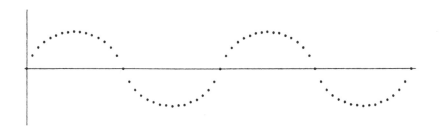

Fig. 1.4.3 - Y=Y+Sin(I*J)/(J*J) for odd terms.

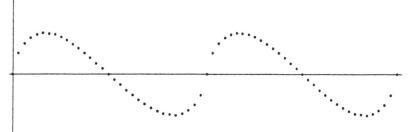

Fig. 1.4.4 - Y=Y+Sin(I*J)/(J*J) for all terms.

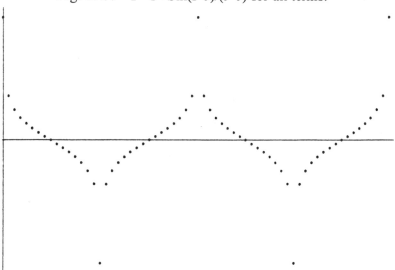

Fig. 1.4.5 - Y=Y+Cos(I*J)/J for odd terms.

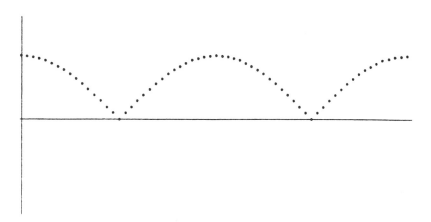

Fig. 1.4.6 - Y=Y-(-1)^J*Cos(J*I)/(4*J*J-1) for all terms.
Initialize Y to Y = 0.5

combinations of amplitudes and phases for the harmonic components, which correspond to an infinite number of time domain
waveshapes; unfortunately, this falls short of proving that *any*
waveshape can be produced by this means.

In any case the above illustration has the cart before the
horse. We are almost always provided with a time domain waveshape for which we must find the equivalent frequency domain
spectrum. It is apparent here that one of the underlying assump-
tions of generalized Fourier Analysis is that time domain signals
must, in fact, have frequency domain equivalents.

1.07.1 FREQUENCY DOMAIN

Figure 1.5 plots the amplitudes of the harmonic components against the harmonic number of the component, displaying

the "spectrum" of a square wave. Now, in accordance with our earlier definition of a function, we recognize that this spectrum is itself a function. The harmonic number (or more commonly the equivalent frequency) represents the independent variable of this function, and the amplitude of the harmonic component represents

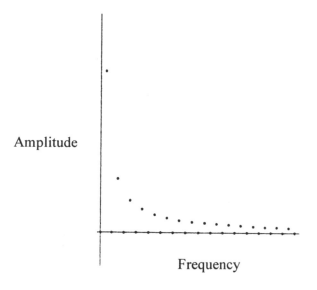

Figure 1.5 - Square Wave Spectrum

the dependent variable. The total interval of the frequencies represents the domain of this new function; consequently, we refer to this function as the frequency domain function. It is this frequency domain function that we seek to obtain with Fourier Analysis (i.e. the transformation from the time domain to the frequency domain).

It should be apparent that the frequency domain function describes the same entity as the time domain function. In the time domain all of the sinusoid components are summed together into the resultant. In the frequency domain, however, we separate out the components and plot the amplitudes (and phases) of the individual sinusoids. It should be absolutely clear, then, that we are looking at the same thing here.

1.07.2 REALITY OF THE FREQUENCY DOMAIN

When first presented with the proposition that all time domain waveshapes are composed of sinusoids, we tend to question the physical reality of the components. We "know" that the time domain signal is the "real" signal and the frequency components are "just another way of analyzing things." Seasoned veterans, however, have no difficulty accepting the sinusoids as completely real. Let us stop here and ask, once and for all, *are the sinusoids real?* Or are they only mathematical gimmicks? Or is this, in fact, a moot question?

The education of electrical engineers, for example, is grounded in the frequency domain. They are taught to *think* in terms of the frequency domain. They are taught to test their circuits by driving the input with a sinusoid while observing the output. By repeating this test for a range of frequencies they determine the *frequency response* of their circuits. As a specific example, they rarely think about audio in the time domain—music is an ever changing kaleidoscope of fundamentals and harmonics. Elsewhere, they learn that modifying the frequency response of a circuit in certain ways will achieve predictable modifications to the time response, e.g. *low pass filtering* will reduce the higher frequency components thereby reducing noise, slowing *rise times*, etc. This sort of experience, coupled with the knowledge that waveforms can be viewed as summations of sinusoids, leads the student into a paradigm that actually prefers the frequency domain. Engineers can always arrive at completely logical and self-consistent conclusions in the frequency domain, and frequently

with much less work than in the time domain. After working in
the field for a few years the notion of frequency domain takes on
a sense of reality for engineers and technicians that others may not
share.

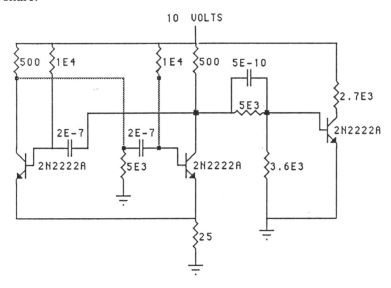

Fig. 1.6 - Astable Multivibrator

Let's look at a concrete example—suppose we build an
astable multivibrator and use it to generate a square wave (actually,
astable multivibrators do not produce waveshapes that are very
"square", so a "buffer" stage is added in the schematic above).
When we view the output waveshape we might justifiably ask,
"where are all the sine waves?" (See Fig. 1.7 below.) On the other
hand, we could *synthesize* a square wave by combining the outputs
of thousands of sine wave generators just as we did with the
computer program several pages back. When we had finished syn-
thesizing this waveform, we would have produced the same thing

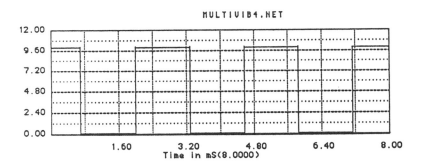

Fig. 1.7 - Output Waveform of Astable Multivibrator

the astable multivibrator produced—a square wave (allowing that our generators produced harmonics that extended beyond the bandwidth of the testing circuitry). If we took some instrument (such as a *wave analyzer* or *spectrum analyzer*) that was capable of measuring the harmonic components of our synthesized wave, we would expect to find each of the sine wave components just

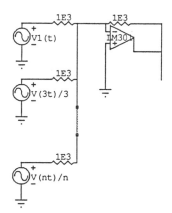

NOTE: Each generator is a sine wave signal source. The frequencies are odd multiples of the "fundamental" V1(t) generator and the amplitudes are inversely proportional to their frequency.

Fig. 1.8 - Square Wave Synthesizer

summed into that wave shape. *But the time domain wave shape of the synthesizer is the same as the output of our multivibrator.* If we use our wave analyzer on the multivibrator output, we will surely find the same components that we found in the synthesized wave, because the two time domain wave shapes are the same. The two are equivalent. A summation of sinusoids is the same thing as the time domain representation of the signal. That is what the examples of Fig. 1.4.x illustrate. A multivibrator may be thought of as a clever device for simultaneously generating a great many sinusoids. The only difference is in our perception—our understanding.

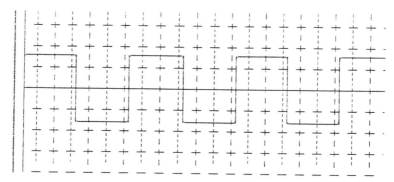

Fig. 1.9 - Synthesizer Waveform

1.08 WHAT IS THE DFT?

The DFT is a procedure, or process, that can analyze the data points of a "digitized" time domain function to determine a series of sinusoids which, when summed together, reproduce the data points of the original function. The resulting Digital Fourier series is a valid expression of the original function, just as the Taylor series examples given in section 1.01 are valid expressions of sines, cosines, etc. It is apparent, however, that the Digital

Fourier Transform is different from the Taylor series, although the exact nature of the difference may still be less than completely obvious. Let's take a moment and focus precisely on some of the differences: the Taylor series, as illustrated in equations (1.4) and (1.5), *evaluate* a specific function at a given argument. The *coefficients* for any specific series are determined once, and thereafter never change. The Taylor series is used in calculators and computers to evaluate a sin, cosine, exponential, etc., etc. for a given argument, and when we use the Taylor series, only the argument changes. Now the DFT is a process used to *determine the coefficients of a trigonometric series* for a given function (i.e. we analyze an arbitrary function to determine the amplitudes (the coefficients) for a series of sinusoids). In contrast to the Taylor series, the *arguments* of a DFT function are fixed and usually remain unchanged; when operating in the frequency domain it is generally the *coefficients* of the transformed function that we modify. Obviously, the Fourier series and the Taylor series have completely different purposes.

What, then, is the purpose of the DFT? A great many procedures, techniques and theorems have been developed to work with functions in the frequency domain. As it turns out, in the frequency domain we may easily perform relatively difficult mathematical techniques like differentiation, integration, or convolution via simple multiplication and division (in some cases this is the *only* way we can perform these operations). At a higher level of problem solving, we can perform minor miracles. We may, of course, examine the frequency spectra of time domain waveshapes, and taking the next obvious step, perform digital filtering. From here it is only a small step to enhance photographic images bringing blurry pictures into sharp focus, but we may continue along this line of development to remove image distortions due to aberrations in the optical system (re: the Hubble telescope). We can do other things that may not be so obvious such as speed up the playback of recorded messages without changing pitch, or convert a television format from 50 frames/sec to 60 frames/sec without speeding up the action. Working in the frequency domain we can perform still more amazing things than these, but the best is certainly yet to come as a new generation of scientists and engineers continue to explore and develop the field of Digital Signal Processing (DSP).

There is a difference between the Taylor and Fourier series that we still may not have made apparent. The terms of the Taylor series are summed to evaluate the function at a single point (i.e at some specific argument). The transformation and inverse transformation of the DFT, on the other hand, involves *all* of the values of the function within the domain of definition. That is, we transform the whole function. When we speak of a function proper, we are not talking about the value of the function at any specific point, but rather, we are talking of the *values of all of its points*. It is one of the little marvels of the DFT that it can transform all of the points of a function, as it were, simultaneously, from the time domain to the frequency domain—and then back to the time domain again via the inverse transform.

1.09 WHAT IS THE FFT?

What the FFT is, of course, is the question we will spend most of this book answering. For the moment though it will be worthwhile to present an analogy which shows clearly what we hope to accomplish. In calculators and computers the approximation of functions such as SIN(X), COS(X), ATN(X), EXP(X), etc., etc., may be obtained by the Taylor series (as we explained previously); but, there is a problem in applying these series directly—they are too slow! They take too long to converge to the accuracy required for most practical work. Use of the Taylor series would be severely limited had not our friends, the mathematicians, figured out the following way to make it run faster:

We observe, as a practical matter, that all of the different series required are of the polynomial form:

$$F(x) = A_o + A_1x + A_2x^2 + A_3x^3 + A_4x^4 + ... \qquad (1.10)$$

where the A_n terms must be substituted into the polynomial for the specific function being evaluated (see eqns 1.4 and 1.5 for examples). The "Horner Scheme" takes advantage of this generalization by solving the polynomial in the following form:

$$F(x) = A_0 + x(A_1 + x(A_2+x(A_3+..+(xA_n)..))) \text{ --- } \qquad (1.11)$$

where we have repeatedly factored x out of the series at each succeeding term. Now, at the *machine language* level of operation, numbers are raised to an integer power by repeated multiplication, and an examination of (1.10) and (1.11) above will show that for an Nth order polynomial this scheme reduces the number of multiplications required from $(N^2+N)/2$ to N. When one considers that N ranges upwards of 30 (for double precision functions), where the Horner Scheme yields execution times an order of magnitude faster, the power of this algorithm becomes apparent.

The above is particularly prophetic in our case. The DFT, although one of the most powerful weapons in the digital signal processing arsenal, suffers from the same malady as the Taylor series described above—when applied to practical problems it tends to bog down—it takes too long to execute. The FFT, in a way that is quite analogous to the Horner Scheme just described, is an algorithm that greatly reduces the number of mathematical operations required to perform a DFT. Unfortunately the FFT is not as easy to explain as the Horner Scheme; although, as we shall see, it is not as difficult as the literature usually makes it out to be either.

1.10 CONCLUSION/ HISTORICAL NOTE

Throughout this chapter we have repeated the proposition that *physically realizable* waveshapes can always be represented as a summation of sine and cosine waves. We have also discussed things such as the nature of "functions", etc., but the summation of sinusoids has obviously been our central theme. *This proposition is the foundation of Fourier Analysis.* The primary purpose of this chapter has been to convey this fundamental idea.

The widespread use of Fourier Analysis implies this proposition is valid; still, when we are presented with a concept whose logical foundations are not readily apparent, our natural curiosity makes us wonder how it came about. Who was the first to discover it, and how did they figure it out? What made someone suspect that all functions could be represented as a series of sinusoids? Early on we saw that the summation of sinusoids could produce complex looking waveshapes. A perceptive soul, recognizing this fact, might well move on to investigate how far this process could be extended, what classes of functions could be evaluated by this method, and how the terms of each such series could be determined.

$$F(x) = A_0 + A_1Cos(x) + A_2Cos(2x) + A_3Cos(3x)+...$$
$$+ B_1Sin(x) + B_2Sin(2*x) + B_3Sin(3x)+.. \qquad (1.11)$$

Daniel Bernoulli, in the 18th century, recognized that

functions could be approximated by a trigonometric series, and
many mathematicians worked with the notion afterward, but it was
Jean Baptiste Joseph Fourier, in the 19th century, who demon-
strated the power of this technique as a practical, problem solving
tool. We might note that this did not bring Fourier immediate
praise and fame, but rather, harsh criticism and professional
frustration. His use of this technique was strongly opposed by no
less a mathematician than Lagrange (and others). Lagrange was
already familiar with trigonometric series, of course, but he also
recognized the peculiarities of their behavior. That trigonometric
series were universally applicable was not at all obvious at that
time.
        The point here is that Fourier did not completely under-
stand the tool he used, nor did he invent it. He had no proof that
trigonometric series could provide universally valid expressions for
all functions. The picture we see here is of brilliant men struggling
with mathematical concepts they cannot quite grasp, and we begin
to realize that the question, "Who invented Fourier Analysis?" is
somewhat naive. There was no single great flash of insight; there
were only many good men working tirelessly to gain understanding. Today no one questions the application of Fourier Analysis
but, in fact, Lagrange was correct: there *are* functions that *cannot*
be transformed by Fourier's method. Fortunately, these functions
involve infinities in ways that never occur in physically realizable
systems, and so, Fourier is also vindicated.

*Books on Fourier Analysis typically have a short historical note
on the role of J.B.J. Fourier in the development of trigonometric series.
Apparently there is a need to deal with how a thing of such marvelous
subtlety could be comprehended by the human mind—how we could
discover such a thing. While the standard reference is J. Hérivel,* Joseph
Fourier, The Man and the Physicist, *Clarendon Press, one of the better
summations is given by R.N. Bracewell in chapter 24 of his text* The
Fourier Transform and its Applications, *McGraw Hill. He also sheds
light on the matter in the section on Fourier series in chapter 10.*

CHAPTER II

FOURIER SERIES AND THE DFT

2.0 INTRODUCTION

It is assumed that most readers will already be familiar with the Fourier series, but a short review is nonetheless in order to re-establish the "mechanics" of the procedure. This material is important since it is the foundation for the rest of this book. In the following, considerations discussed in the previous chapter are assumed as given.

2.1 MECHANICS OF THE FOURIER SERIES

You may skip section 2.1 with no loss of continuity. It is the only section in this book that employs Calculus. The Fourier series is a trigonometric series F(f) by which we may approximate some arbitrary function f(t). Specifically, F(f) is the series:

$$F(f) = A_o + A_1 Cos(t) + B_1 Sin(t) + A_2 Cos(2t) + B_2 Sin(2t) + ...$$
$$...+ A_n Cos(nt) + B_n Sin(nt) \quad \text{----------------------} \quad (2.1)$$

and, in the limit, as n (i.e. the number of terms) approaches infinity:

$$F(f) = f(t) \quad \text{-----------------------------------} \quad (2.2)$$

The problem we face in Fourier Analysis, of course, is to find the coefficients of the frequency domain sinusoids (i.e. the values of A_o, A_1,...A_n, and B_1,...B_n) which make eqn. (2.2) true.

Finding A_o is easy—if we integrate F(f) (i.e. eqn. 2.1) from 0 to 2π, *all* sinusoid terms yield zero so that only the A_o term is left:

$$\int_0^{2\pi} F(f)\ dt = A_o 2\pi \quad \text{--------------------} \quad (2.3)$$

From eqn.(2.2) and the following condition:

$$1/2\pi \quad \int_0^{2\pi} f(t)\ dt = \text{mean value} \quad \text{------------} \quad (2.4)$$

it follows that:

$$A_o = 1/2\pi \int_0^{2\pi} f(t)\ dt = \text{mean value} \quad \text{-----} \quad (2.5)$$

Next, if we multiply both sides of eqn.(2.1) by cos(t) and integrate from 0 to 2π, the only non-zero term will be:

$$\int_0^{2\pi} F(f)\cos(t)\ dt\ =\ \int_0^{2\pi} A_1 \cos^2(t)\ dt\ =\ \pi A_1 \quad --- \quad (2.6)$$

This results from the fact that:

$$\int_0^{2\pi} \cos(rx)\cos(qx)\ dx = 0 \quad ---------------- \quad (2.7.1)$$

$$\int_0^{2\pi} \cos(rx)\sin(px)\ dx = 0 \quad ---------------- \quad (2.7.2)$$

$$\int_0^{2\pi} \sin(rx)\sin(qx)\ dx = 0 \quad ---------------- \quad (2.7.3)$$

Where: r, q, and p are integers and $r \neq q$

From eqns.(2.2) and (2.6) then we may evaluate A_1:

$$A_1 =\ 1/\pi\ \int_0^{2\pi} f(t)\cos(t)\ dt \quad ----------- \quad (2.8)$$

From the same argument, if we multiply eqn.(2.1) by Sin(t) and integrate from 0 to 2π, the only non-zero term will be:

$$\int_0^{2\pi} B_1 Sin^2(t)\, dt = \pi B_1 \qquad \text{-----------------} \qquad (2.9)$$

We may therefore evaluate B_1:

$$B_1 = 1/\pi \int_0^{2\pi} f(t)Sin(t)\, dt \qquad \text{----------} \qquad (2.10)$$

If we continue through the other terms of eqn. (2.1) we will find
that the procedure for determining the A and B coefficients may be
summarized by the following:

$$A_o = 1/2\pi \int_0^{2\pi} f(t)\, dt \qquad \text{-----------------} \qquad (2.11A)$$

$$A_k = 1/\pi \int_0^{2\pi} f(t)\cos(kt)\, dt \qquad \text{----------} \qquad (2.11B)$$

$$B_k = 1/\pi \int_0^{2\pi} f(t)\sin(kt)\, dt \qquad \text{----------} \qquad (2.11C)$$

With: $k = 1, 2, 3, \ldots, n$
        $n = $ number of terms included in the series.

As n approaches infinity, we must necessarily include all possible
sinusoidal (by sinusoidal we imply both sine and cosine) components, and $F(f)$ converges to $f(t)$.

COMMENTARY

We should take the time to point out a few things about the above derivation. Our starting equations (2.1) and (2.2) simply make the statement that a given arbitrary function f(t) may be considered to be a summation of sinusoids as explained in the previous chapter. It is well known that functions exist for which this condition is untrue; fortunately, it *is* true for all physically realizable systems.

Equations (2.8) and (2.10) express the mathematical operation that is the heart of the Fourier series; the individual sinusoids of a composite wave can be "detected" by multiplying through with unit sinusoids and finding the mean value of the resultant. This process is all that the DFT (and FFT) does.

The reader should understand that the relationships expressed in equations (2.7.1) through (2.7.3) (i.e. the orthogonality relationships) are imperative to the proper operation of this algorithm; furthermore, these equations are true only when evaluated over an integer number of cycles. In practice the Fourier series, DFT, and FFT force this condition for any time domain T_o by restricting the arguments of the sinusoids to integer multiples of $2\pi Nt/T_o$ (where N is an integer).

In addition to these comments, the DFT deals with arrays of discrete, digital data. There are no linear, continuous curves in a computer. We will spend the rest of this chapter delving into how we apply the mathematical process described so neatly above to the digitized data we process inside a computer.

2.2.0 MECHANICS OF THE DFT

The DFT is an application of Fourier Analysis to discrete (i.e. digital) data. Our objective in this chapter is to find out what makes the DFT work—and why. At considerable risk to personal reputation, we will employ only simple, direct illustrations. It would be safer to brandish the standard confusion and abstruse mathematics; but then, there are plenty of books already on the market to fulfill that requirement. We will start from the concepts covered in the previous chapter and develop our own process for extracting harmonic components from arbitrary waveforms.

2.2.1 THE OBJECTIVE OF THE DFT PROCESS

We saw in chapter 1 that sinusoids could be summed together to create common waveforms. Here we consider the reverse of that process. That is, given some arbitrary waveshape, we try to break it down into its component sinusoids.

Let's talk about this for a moment. When we are presented with a composite waveshape, all of the sinusoids are "mixed together," sort of like the ingredients of a cake—our question is, "how does one go about separating them?" We already know, of course, that it *is* possible to separate them (not the ingredients of cakes—the components of composite waveforms), but let's forget for a moment that we know about Fourier Analysis...

2.2.2 REQUIREMENTS FOR A DFT PROCEDURE

There are two, perhaps three, requirements for a method to "separate out" the components of a composite wave:

1) First, we require a process that can isolate, or "detect", any single harmonic component within a complex waveshape.

2) To be useful quantitatively, it will have to be capable of measuring the amplitude and phase of each harmonic component.

These, of course, are the primary requirements for our procedure; but, there is another requirement that is implied by these first two:

3) We must show that, while measuring any harmonic component of a composite wave, our process ignores all of the other harmonic components (i.e. it must not include any part of the other harmonics). In other words, our procedure must measure the *correct* amplitude and phase of the *individual* harmonics. Very well then, let's see how well the DFT fulfills these requirements.

2.2.3 THE MECHANISM OF THE DFT

Let's begin with a typical digitized sine wave as shown in Figure 2.1 below. The X axis represents time; the Y axis represents volts (which, in turn, may represent light intensity, or pressure, or etc.) It has a peak amplitude of 2.0 volts, but its average value is obviously zero. The digitized numbers (i.e. our function) are stored in a nice, neat, "array" within the computer (see Table 2.1). The interval from the beginning to end of this array is the domain (see first column Table 2.1).

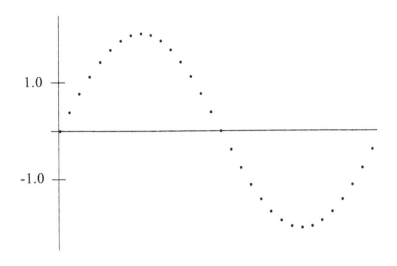

Figure 2.1 - Digitized Sinusoid

    Now, according to Fourier, the way to detect and measure a sine wave component within a complex waveshape is to multiply through by a unit amplitude sine wave (of the identical frequency), and then find the *average value* of the resultant. This is the fundamental concept behind Fourier Analysis and consequently we will review it in detail. First, we create a *unit amplitude* sine wave (see Fig. 2.2). To multiply our digitized waveform by a unit sine wave we multiply each point of the given function by the corresponding point from the unit sinusoid function (apparently the two

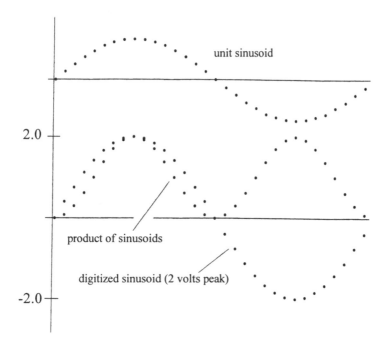

unit sinusoid

product of sinusoids

digitized sinusoid (2 volts peak)

2.0

-2.0

Figure 2.2 - Fourier Mechanism

functions must have the same number of data points, corresponding domains, etc.) This process (and the result) is shown in Figure
2.2 and Table 2.1. The reason this works, of course, is that both
sine waves are positive at the same time (yielding positive
products), and both are negative at the same time (still yielding
positive products), so that the products of these two functions will
have a positive average value. Since the average value of a
sinusoid is normally zero, this process has "rectified" or "detected"
the original digitized sine wave of Fig. 2.1. The sum of all of the
products is 16 (see Table 2.1 below), and since there are 16 data
points in the array, the average value is 1.0.

T	Sin (X)	* (2*Sin (X))	= 2*Sin²(X)
0.00000	0.00000	0.00000	0.00000
0.06250	0.38268	0.76537	0.29289
0.12500	0.70711	1.41421	1.00000
0.18750	0.92388	1.84776	1.70711
0.25000	1.00000	2.00000	2.00000
0.31250	0.92388	1.84776	1.70711
0.37500	0.70711	1.41421	1.00000
0.43750	0.38268	0.76537	0.29289
0.50000	0.00000	0.00000	0.00000
0.56250	-0.38268	-0.76537	0.29289
0.62500	-0.70711	-1.41421	1.00000
0.68750	-0.92388	-1.84776	1.70711
0.75000	-1.00000	-2.00000	2.00000
0.81250	-0.92388	-1.84776	1.70711
0.87500	-0.70711	-1.41421	1.00000
0.93750	-0.38268	-0.76537	0.29289
Totals =	0.00000	0.00000	16.00000

Average Value = 16.0/16 = 1.00000

Table 2.1

Note that the *average* amplitude of the right hand column is only half of the *peak* amplitude of the input function (3rd column). We may show that the average value obtained by the above procedure will always be half of the original input amplitude as follows:

The input function is generated from the equation shown on the following page:

$$F_1(T) = A \sin(2\pi t) \quad \text{-----------------------} \qquad (2.12)$$

Where: t = Time
A = Peak amplitude
NOTE: A frequency (or period) of unity is implied here

We simplify by replacing the argument $(2\pi t)$ with X:

$$F(X) = A \, Sin(X) \quad \text{----------------} \qquad (2.13)$$

Multiplying through by a sine wave of unit amplitude:

$$F(X) \, Sin(X) \quad = A \, Sin(X)Sin(X) \quad \text{----} \qquad (2.14)$$
$$= A \, Sin^2(X) \quad \text{---------} \qquad (2.14A)$$

However, from the trigonometric identity:

$$Sin^2(X) = 1/2 - Cos(2X)/2 \quad \text{-------} \qquad (2.15)$$

which we substitute into (2.14A):

$$F(X) \, Sin(X) \quad = A \, (1/2 - Cos(2X)/2)$$
$$= A/2 - ACos(2X)/2 \quad \text{--} \qquad (2.16)$$

The second term of (2.16) (i.e. A Cos(2X)/2) describes a sinusoid so that its *average* value will be zero over any number of full cycles; it follows that the average value of eqn. 2.16 (over any integer multiple of 2π radians) will be A/2 (see figure 2.3).

This result is more or less obvious from an inspection of Figs. 2.2 and 2.3. It is apparent that the maximum value will occur at the peaks of the two sinusoids, where the unit sinusoid has an

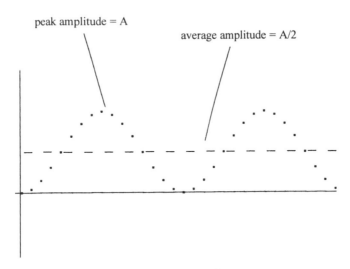

peak amplitude = A

average amplitude = A/2

Figure 2.3 - A Sin² Wave

amplitude of 1.0 and the product of the two functions is A. The minimum value will occur when the sinusoids are passing through zero. From the symmetry of Fig. 2.3 it is apparent that the average value must be A/2.

This process of detecting or rectifying sinusoids, then, has the characteristic of yielding only half the amplitude of the actual component. This presents no major problem though, as we can simply multiply all of the results by two, or use some other technique to correct for this phenomenon.

2.2.4 THE COSINE COMPONENT

It is obvious that this scheme will work for any harmonic component; we need only change the frequency of the *unit amplitude sine wave* to match the frequency of the harmonic component being detected. This same scheme will work for the cosine components if we replace the unit sine function with a unit cosine function.

The component we want to detect is given by:

$$F_2(T) = A \cos(2\pi t) = A \cos(X) \quad ----- \qquad (2.17)$$

Where: All terms are as in (2.12) and (2.13) above.

Multiplying through by a cosine wave of unit amplitude:

$$\begin{aligned} F(X) \cos(X) &= A \cos(X) \cos(X) \quad -- & (2.18) \\ &= A \cos^2(X) \quad ---------- & (2.18A) \end{aligned}$$

From the identity:

$$\cos^2(X) = 1/2 + \cos(2x)/2 \quad ----------- \qquad (2.19)$$

which we substitute into (2.18A):

$$\begin{aligned} F(X) \cos(X) &= A\,(1/2 + \cos(2X)/2) \\ &= A/2 + A \cos(2X)/2 \quad --- & (2.20) \end{aligned}$$

Again, the second term will have a zero average value while the first term is one half the input amplitude. Note carefully in the above developments that, to produce a workable scheme, we must design our system such that we always average over full cycles of the sinusoids.

2.2.5 HARMONICS WITHIN COMPOSITE WAVEFORMS

Next question: "What about the *other* harmonics in composite waveforms? The scheme described above undoubtedly works when we are dealing with a single sinusoid, but when we are dealing with a composite of many harmonics, how do we know that *all* of the other harmonics are *completely* ignored? You will recall that our third requirement for a method to extract sinusoidal components from a composite wave was that the process ignore all but the sinusoid being analyzed. Technically, this condition is known as orthogonality.

2.2.6 ORTHOGONALITY

What, exactly, does the term orthogonality imply? Two straight lines are said to be orthogonal if they intersect at right angles; two curved lines are orthogonal if their tangents form right angles at the point of intersection. Consider this: the "scaler product", or "dot product" between two vectors is defined as follows:

$$\mathbf{A \cdot B} = |A| \, |B| \, \mathrm{Cos}\,(\emptyset) \text{ ------------------} \qquad (2.21)$$

$$|A| = \text{magnitude of vector } \mathbf{A}, \text{ etc.}$$

As the angle $\emptyset$ between the two vectors approaches $\pm$ 90 degrees, Cos $\emptyset$ approaches zero, and the dot product approaches zero. It is apparent, then, that a zero dot product between any two *finite* vectors implies orthogonality. It is apparent that *zero magnitude*

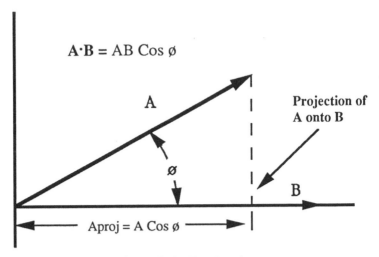

A·B = AB Cos ∅

A

Projection of
A onto B

∅

B

Aproj = A Cos ∅

Figure 2.4 - Dot Product

vectors will always yield a zero dot product regardless of the angle (in fact, the notion of angle begs for definition here). In practice we effectively *define* vectors of zero magnitude as orthogonal to *all* other vectors, and may therefore use a zero resultant from equation (2.21) as the operative definition for orthogonality.

DEFINITION 1:

If **A·B** = 0 then **A** and **B** are orthogonal.

Note that, by this definition, zero magnitude vectors are orthogonal to all other vectors.

The definition of orthogonality between whole functions derives from an argument not completely dissimilar to the above. To illustrate, we will first use equation (2.21) to generate a function F(ϕ). We will do this with a specific example where **A** is

a unit vector lying along the X axis and **B** is a unit vector in the
direction of ϕ (where ϕ takes on a set of values over the domain of
0 to 2π). Table 2.2 shows this function explicitly—note that,
according to our definition above, the two vectors are orthogonal
at only two points (i.e. $\phi = \pi/2$ and $3\pi/2$).

ϕ	$F(\phi)$
0	1.000
$\pi/8$	0.924
$\pi/4$	0.785
$3\pi/8$	0.383
$\pi/2$	0.000
$5\pi/8$	-0.383
$3\pi/4$	-0.785
$7\pi/8$	-0.924
π	-1.000
$9\pi/8$	-0.924
$5\pi/4$	-0.785
$11\pi/8$	-0.383
$3\pi/2$	0.000
$13\pi/8$	0.383
$7\pi/4$	0.785
$15\pi/8$	0.924

Table 2.2 - $F(\phi) = |A| \, |B| \cos(\phi)$

But our concern is not for individual values of a function;
our concern is for orthogonality between whole functions. To
investigate this situation we need a second function, $G(\phi)$, which
we will define as follows:

$$G(\phi) = |A| \, |B| \cos(\phi + \pi/2) \quad \text{----------------} \quad (2.22)$$

Let's talk about these two functions, F(ϕ) and G(ϕ), for a moment. For both functions vector **A** lies along the X axis, while **B** "rotates" about the origin as a function of ϕ. The difference between these two functions is that the argument of G(ϕ) is advanced by $\pi/2$ (i.e. 90 degrees) so that the vector **B** of G(ϕ) will be orthogonal to **B** in

ϕ	F(ϕ)	G(ϕ)	F(ϕ)G(ϕ)
0	1.000	0.000	0.000
$\pi/8$	0.924	-0.383	-0.354
$\pi/4$	0.785	-0.785	-0.616
$3\pi/8$	0.383	-0.924	-0.354
$\pi/2$	0.000	-1.000	0.000
$5\pi/8$	-0.383	-0.924	0.354
$3\pi/4$	-0.785	-0.785	0.616
$7\pi/8$	-0.924	-0.383	0.354
π	-1.000	0.000	0.000
$9\pi/8$	-0.924	0.383	-0.354
$5\pi/4$	-0.785	0.785	-0.616
$11\pi/8$	-0.383	0.924	-0.354
$3\pi/2$	0.000	1.000	0.000
$13\pi/8$	0.383	0.924	0.354
$7\pi/4$	0.785	0.785	0.616
$15\pi/8$	0.924	0.383	0.354
		Sum Total	0.000

Table 2.3 - F(ϕ)G(ϕ)

F(ϕ) for all values of ϕ. In other words, the vectors which actually generate these two functions are orthogonal at all points of the functions.

Now the two functions created here are not vectors; being scalar products of vectors, they are scalars. Therefore, the definition of orthogonality given above is inappropriate. Still, these two

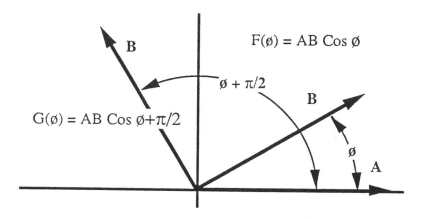

Fig. 2.5 - Orthogonal Functions F(ϕ) and G(ϕ)

functions were created by orthogonal vectors and we would like to find out if there is some vestige—some latent characteristic—that can still expose the orthogonality relationship. In fact, there is. If we multiply these two functions point by point, and then sum all of the individual products, the result will be zero (see Table 2.3). You may recognize that this process is essentially the process we used to detect sinusoids back in section 2.2.3. If so, you may also begin to grasp the connection we are trying to make here, but bear with me for a moment longer. Let's generalize the above results as follows:

$$G(\phi) = |A| \, |B| \cos(\phi+\Omega) \quad \text{----------------} \quad (2.23)$$

If we replace the $\pi/2$ in G(ϕ) by the variable Ω, and repeat the test for orthogonality using various values of Ω (see Table 2.4), we see that a zero resultant is achieved only when $\Omega = \pi/2$, indicating orthogonality between the vectors which generate the functions.

$$
\begin{array}{cc}
\Omega & \sum_{i=0}^{N} F(\phi)_i G(\phi)_i
\end{array}
$$

Ω	$\sum_{i=0}^{N} F(\phi)_i G(\phi)_i$
0	8.000
$\pi/8$	7.391
$\pi/4$	5.657
$3\pi/8$	3.061
$\pi/2$	0.000
$5\pi/8$	-3.061
$3\pi/4$	-5.657
$7\pi/8$	-7.391
π	-8.000

Table 2.4 - Orthogonality Test

So then, this process does indeed detect "orthogonality" between our functions. It is not necessary to prove this relationship for all cases, for at this point we simply adopt this procedure as our operative definition for orthogonality between functions. If we compute products between all of the points of the functions, and the summation of these products is zero, then the functions are orthogonal. Newcomers sometimes find this remarkable, for by this definition, we no longer care how the functions were generated. Nor do we ask that the functions be traceable to geometrically orthogonal origins in any sense. The *only* requirement is that the summation of the products of corresponding points between the two functions be zero.

DEFINITION 2:

If $\sum_{i=0}^{N} F(\phi)_i G(\phi)_i = 0$ then $F(\phi)$ and $G(\phi)$ are orthogonal

Orthogonality then is a condition which applies directly to the process we use as the Fourier mechanism. Some functions will be orthogonal (i.e. they will always give a zero resultant when we perform the process described above) and others will not. Obviously the example of section 2.2.3 (i.e. two sine waves of identical frequency) *does not* illustrate orthogonal functions.

The question here, however, is whether two sinusoids of *different, integer multiple frequencies* represent orthogonal functions. This, of course, is an imperative condition. If they *are* orthogonal, components that are not being analyzed will contribute zero to the resultant—if they *are not* orthogonal we have BIG problems. Let's see how this works out.

As before, we start with a digitized sine wave but this time we multiply through with a sine wave of twice the frequency (Fig. 2.6 below). By symmetry it is more or less apparent that this yields an average value of zero. Apparently we have orthogonal functions here, but we need a demonstration for the general case of *any* integer multiple frequency.

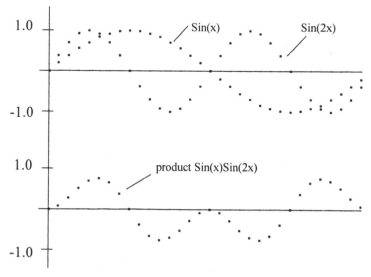

Figure 2.6 - Sin(x)Sin(2x)

Starting with the identity:

$$Sin(A)Sin(B) = \frac{Cos(A-B) - Cos(A+B)}{2} \quad --- \quad (2.24)$$

if we let A and B represent arguments of:

$$A = 2\pi t \quad \text{and} \quad B = NA = 2N\pi t$$
$$N = 1,2,3,4,... \text{ (i.e. N takes on integer values)}$$

eqn. (2.24) becomes:

$$Sin(A)Sin(NA) = \frac{Cos(A(1-N)) - Cos(A(1+N))}{2} \quad (2.25)$$

It is interesting to note when N=1, the term $Cos(A(1-1))$ yields a value of 1 regardless of the value of A (i.e. all values of $2\pi T$), and eqn.(2.25) reduces to:

$$Sin(A)Sin(A) = \frac{1 - Cos(2A)}{2} \quad ------------ \quad (2.26)$$

which is the same equation as (2.15) above. The term $Cos(2A)$ generates a sinusoid as A varies from 0 to 2π and must therefore have a zero average value.

     If we consider all other positive values of N in eqn.(2.25) we will always obtain a non-zero value for the argument of both terms. Consequently, both terms on the right in eqn.(2.25) will generate sinusoids, which guarantees an average value of zero for the function (averaged over any number of full cycles).

The second case we must examine is for two *cosine* waves of different frequencies. As before we start with an examination of the trigonometric identity:

$$Cos(A)Cos(B) = \frac{Cos(A+B) + Cos(A-B)}{2} ----- \qquad (2.27)$$

When $B = NA$ then (2.27) becomes:

$$= \frac{Cos(A(1+N)) - Cos(A(1-N))}{2} -- \qquad (2.27A)$$

which shows that these functions are also orthogonal except when $N=1$ and the arguments (i.e. the frequencies) are identical.

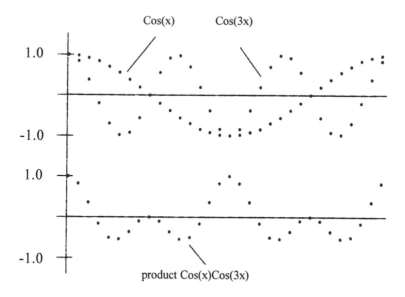

Figure 2.7 - Cos(x)Cos(3x)

Finally, we must examine the relationship between the cosine and sine functions for orthogonality. This can be shown by the following identity:

$$Sin(A)Cos(B) = \frac{Sin(A+B) + Sin(A-B)}{2} \quad ---- \quad (2.28)$$

and when B = NA:

$$= \frac{Sin(A(1+N)) + Sin(A(1-N))}{2} \quad ------ \quad (2.28A)$$

In this case it makes no difference whether the N = 1 or not; if the argument is zero the value of the sine function is likewise zero; if the argument is multiplied by an integer the function will trace out an integer number of cycles as A varies from 0 to 2π. In no case will the average value of these functions be other than zero—*they are always orthogonal.*

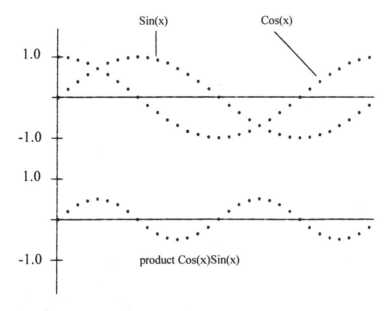

Figure 2.8 - Cos(x)Sin(x)

2.2.6 THE DFT/FOURIER MECHANISM

Finally, we must consider all of this together. We know that the composite waveform is generated by summing in harmonic components:

$$F(f) = A_0 + A_1Cos(t) + B_1Sin(t) + A_2Cos(2t) + B_2Sin(2t) + ...$$
$$+ A_nCos(nt) + B_nSin(nt) \text{ --------------}\qquad (2.29)$$

If we multiply this composite function by $Sin(Kt)$ (or, alternatively, $Cos(Kt)$), where K is an integer, we will create the following terms on the right hand side of the equation:

$$A_0Sin(Kt) + A_1Sin(t)Sin(Kt) + B_1Cos(t)Sin(Kt) + ...$$
$$+ A_kSin^2(Kt) + B_kCos(Kt)Sin(Kt) + ...$$
$$+ A_nSin(nt)Sin(Kt) + B_nCos(nt)Sin(Kt)\qquad (2.30)$$

Treating each of these terms as individual functions, if the argument (Kt) equals the argument of the sinusoid it multiplies, that component will be "rectified." Otherwise, the component will not be rectified. From what we have shown above, two sinusoids of $\pm \pi/2$ phase relationship (i.e. Sine/Cosine), or integer multiple frequency, represent orthogonal functions. As such, when summed over all values within the domain of definition, they will all yield a zero resultant (regardless of whether they are handled as individual terms or combined into a composite waveform). That, of course, is precisely what we demanded of a procedure to isolate the harmonic components of an arbitrary waveform. The examples of the next chapter will illustrate the practical reality of these relationships. Since a computer can do little more than simple arithmetic on the input data, computer examples have a way of removing any reasonable question about the validity of theoretical developments.

CHAPTER III

THE DIGITAL FOURIER TRANSFORM ALGORITHM

3.0 INTRODUCTION

The DFT is a simple algorithm. It consists of stepping through the digitized data points of the input function, multiplying each point by sine and cosine functions as you go along, and summing the resulting products into accumulators (one for the sine component and another for the cosine). When we have processed every data point in this manner, we divide the accumulators (i.e. the sum-totals of the preceding process) by the number of data points. The resulting quantities are the average values for the sine and cosine components at the frequency being investigated as we described in the preceding chapter. We must repeat this process for all integer multiple frequencies up to the frequency that is equal to the sampling rate minus 1 (i.e. twice the Nyquest frequency minus 1), and the job is done.

In this chapter we will examine a program that performs the DFT. We will walk through this first program step by step, describing each operation explicitly.

3.1 THE DFT COMPUTER PROGRAM

In the program presented below a "time domain" function is generated (16 data points) by summing together the first 8 harmonic components of the classic "triangle wave." This time domain data is stored in an *array* Y(n), and then analyzed as described above. In this program we use programming and data structuring features common to all higher level languages, *viz.* the data is stored in arrays and the execution of the program takes place via subroutines. Each subroutine works on the data arrays, performing a specific task. This allows the main body of the program (i.e. lines 20 through 80) to operate at a high level, executing the necessary tasks (i.e. the subroutines) in a logical order. Lets begin by looking at the whole program. As you can see, everything is controlled between lines 20 and 60.

```
6  REM  ******************************************
8  REM  *** (DFT3.1) GENERATE/ANALYZE WAVEFORM ***
10 REM  ******************************************
12 PI=3.141592653589793#:P2=2*PI:K1=PI/8:K2=1/PI
14 DIM Y(16),FC(16),FS(16),KC(16),KS(16)
16 CLS:FOR J=0 TO 16:FC(J)=0:FS(J)=0:NEXT
20 GOSUB 108: REM - PRINT COLUMN HEADINGS
30 GOSUB 120: REM - GENERATE FUNCTION Y(X)
40 GOSUB 200: REM - PERFORM DFT
60 GOSUB 140: REM - PRINT OUT FINAL VALUES
70 PRINT:PRINT "MORE (Y/N)? ";
72 A$ = INKEY$:IF A$="" THEN 72
74 PRINT A$:IF A$ = "Y" THEN 16
80 END
```

```
100 REM *****************************************
102 REM *        PROGRAM SUBROUTINES        *
104 REM *****************************************
106 REM *      PRINT COLUMN HEADINGS        *
107 REM *****************************************
108 PRINT:PRINT
110 PRINT "FREQ   F(COS)    F(SIN)    Y(COS)    Y(SIN)"
112 PRINT
114 RETURN
118 REM ******************************
120 REM *** GENERATE FUNCTION Y(X) ***
121 REM ******************************
122 FOR I = 0 TO 15:K3=I*K1
124 Y(I) = COS(K3)+COS(3*K3)/(9)+COS(5*K3)/(25)+COS(7*K3)/49
126 NEXT
128 FOR I=1 TO 7 STEP 2: KC(I)=1/I^2:NEXT
130 RETURN
132 REM ******************************
138 REM *        PRINT OUTPUT         *
139 REM ******************************
140 FOR Z=0 TO 15
142 PRINT Z;"   ";
144 PRINT USING "##.#####_   ";FC(Z),FS(Z),KC(Z),KS(Z)
146 NEXT Z
148 RETURN
200 REM ***************************
202 REM * SOLVE FOR COMPONENTS *
204 REM ***************************
206 FOR J=0 TO 15:REM SOLVE EQNS FOR EACH FREQUENCY
208 FOR I = 0 TO 15:REM MULTIPLY AND SUM EACH DATA POINT
210 FC(J)=FC(J)+Y(I)*COS(J*I*K1):FS(J)=FS(J)+Y(I)*SIN(J*I*K1)
212 NEXT I
214 FC(J)=FC(J)/16: FS(J)=FS(J)/16:REM FIND MEAN VALUE
216 NEXT J
218 RETURN
```

Figure 3.1

Now let's dissect this program and its routines to see how things really get done. At the beginning of the program (line 12) we define the frequently used constants of PI, 2*PI, PI/8, and 1/PI (we will duplicate each section of the program as we go along so that you don't have to flip pages). At line 14 we "DIMension" (i.e. define the size) of the arrays to be used in the program. Array Y(16) will store the 16 data points of the time domain function to be analyzed, while FC(16) and FS(16) will hold the 16 derived amplitudes of the Fourier *cosine* and *sine* components. Similarly,

```
6 REM   *****************************************
8 REM   *** (DFT3.1) GENERATE/ANALYZE WAVEFORM ***
10 REM  *****************************************
12 PI=3.141592653589793#:P2=2*PI:K1=PI/8:K2=1/PI
14 DIM Y(16),FC(16),FS(16),KC(16),KS(16)
16 CLS:FOR J=0 TO 16:FC(J)=0:FS(J)=0:NEXT
```

KC(16) and KS(16) will hold the amplitudes of the sinusoids used to generate the input function (these are saved for comparison to the derived components). Having completed this preliminary work, line 16 clears the screen with a CLS statement, and then initializes the arrays FC(J) and FS(J) by placing a zero in every location. Note that the array proper is the FC() designation and that J only provides a convenient variable to specify the location within the array. We may use any variable (or constant) at any time to specify locations within arrays. The data stored at those locations will be unaffected.

This brings us to the main program (lines 20 through 60), which accomplishes the high level objectives. When the program

```
20 GOSUB 108: REM - PRINT COLUMN HEADINGS
30 GOSUB 120: REM - GENERATE FUNCTION Y(X)
40 GOSUB 200: REM - PERFORM DFT
60 GOSUB 140: REM - PRINT OUT FINAL VALUES
70 PRINT:PRINT "MORE (Y/N)? ";
72 A$ = INKEY$:IF A$="" THEN 72
74 PRINT A$:IF A$ = "Y" THEN 16
80 END
```

comes to the GOSUB instruction at line 20 it will "jump down" to line 108, and so will we. This subroutine prints the column headings. In addition to printing out the amplitudes of the sine and

```
106 REM  *       PRINT COLUMN HEADINGS          *
108 PRINT:PRINT
110 PRINT "FREQ    F(COS)   F(SIN)    Y(COS)   Y(SIN)"
112 PRINT
114 RETURN
```

cosine components (as do most Fourier analysis programs), in this program we also print out the amplitude of the components which were used to generate the input function [i.e. Y(COS)   Y(SIN)]. This allows a direct comparison of output to input and tells us how well the analysis scheme is working. Lines 108 through 112 print this heading and then, at line 114, we encounter a RETURN statement which sends program control back to the instruction *following* the line 20 GOSUB 108 statement (i.e program execution jumps back to line 30).

Line 30 jumps us down to the subroutine located at line 120, which generates the input function. Line 120 is a REMark statement telling us this is where we generate the *time domain* input function Y(X), which we will do by summing the harmonic components known to construct a "triangle wave." At line 122 we set up a loop that steps "I" from 0 to 15 (the variable *I* will count

```
120 REM *** GENERATE FUNCTION Y(X) ***
122 FOR I = 0 TO 15:K3=I*K1
124 Y(I) = COS(K3)+COS(3*K3)/(9)+COS(5*K3)/(25)+COS(7*K3)/49
126 NEXT
128 FOR I=1 TO 7 STEP 2: KC(I)=1/I^2:NEXT
130 RETURN
```

the 16 data points of our triangle wave function—note that K3 is computed each time through the loop (K1 is defined back on line 12 as PI/8). Line 124 is the business end of this routine, it sums the *odd* cosine components (with amplitudes inversely proportional to the square of their frequencies) into each point of array Y(I). Since there are 16 points in the data array we can have a maximum of 8 harmonic components (there must be a minimum of two data points for each "cycle" of the Nyquest frequency)[1]. At line 126 the NEXT statement sends us back through the loop again, until we have stepped through the 2*PI radians of a full cycle of the fundamental. At line 128 we have inserted a loop which puts $1/N^2$ into the odd cosine terms of the KC(I) array (which is, in fact, the amplitudes of the sine waves we used to generate this function). Having done all this, we have completed the generation of our input function, and now RETURN (line 130) to the main program (i.e. to line 40).

```
40 GOSUB 200: REM - PERFORM DFT
```

We are now ready to perform a Fourier Transform of the time domain function in array Y(X). From line 40 we GOSUB to

---

1. Note that only 8 harmonics are used to generate this function (in fact that is all the Nyquest Sampling Theorem will allow), but there are 16 frequencies derived in the DFT. We will discuss this in detail later.

line 206 where we set up a loop. This loop will handle everything
that must be done at each of the harmonic frequencies (in this case
the frequency is designated by J). We must perform a multiplication by cosine and sine at each point of the data array (for the
frequency being worked on) and sum the results into the location
of the FC(J) and FS(J). Line 208 sets up a *nested loop* which will

```
200 REM ***************************
202 REM *  SOLVE FOR COMPONENTS  *
204 REM ***************************
206 FOR J=0 TO 15:REM SOLVE EQNS FOR EACH FREQUENCY
208 FOR I = 0 TO 15:REM MULTIPLY AND SUM EACH DATA POINT
210 FC(J)=FC(J)+Y(I)*COS(J*I*K1):FS(J)=FS(J)+Y(I)*SIN(J*I*K1)
212 NEXT I
214 FC(J)=FC(J)/16: FS(J)=FS(J)/16:REM FIND MEAN VALUE
216 NEXT J
218 RETURN
```

step I from 0 to 15. Note that, just as J indicates the frequency, I
indicates the data point in the input function array. Line 210 sums
into FC(J) the product of the data point at Y(I) multiplied by the
COS(K1*I*J). We are multiplying the Ith data point by the Cosine
of: K1 (i.e. PI/8) multiplied by I (which yields the number of
radians along the fundamental that this data point lies) and then
multiplied by the frequency of the component being extracted (i.e.
J), which yields the correct number of radians for that particular
harmonic. In this same line the "sine term" is also found and
summed into FS(J). At line 212 we encounter the NEXT I
statement, jump back to line 208 and repeat this operation for the
next data point. When we have stepped through the 16 points of
the data array, we move down to line 214 and divide both of these
summations by 16 to obtain the average value. At line 216 we
jump back to line 206 and perform the whole routine over for the

next harmonic. We continue this process until we have analyzed all 16 frequencies (the constant, or "D.C." component, is occasionally referred to as the "zeroth" frequency).

```
60 GOSUB 140: REM - PRINT OUT FINAL VALUES
```

Having completed our Fourier analysis, we then return to line 60 where we jump down to the "PRINT OUTPUT" subroutine located at line 140. We set up a loop counter Z which counts from 0 to 15 (corresponding to the frequencies analyzed) and, in fact, at line 142, we print Z under the column heading "FREQ". Let's make note of a few things that happen here:

```
138 REM *       PRINT OUTPUT        *
140 FOR Z=0 TO 15
142 PRINT Z;"   ";
144 PRINT USING "##.#####_    ";FC(Z),FS(Z),KC(Z),KS(Z)
146 NEXT Z
148 RETURN
```

1) A semicolon separates the PRINT Z and the "   ". This causes them both to be printed on the same line.

2) The "   "; simply causes a space to be printed between the frequency column and the following data (note that another semicolon is used so that the next PRINT statement will still be printed on the same line).

Line 144 then prints out the relevant data with a PRINT USING statement. Line 146 causes the program to go back and print out the next line of data with a NEXT Z.

When the data for all 16 frequencies has been printed we return to the main program (line 70) and ask if "MORE (Y/N)" is desired . Line 72 looks for an input from the keyboard and assigns

```
70 PRINT:PRINT "MORE (Y/N)? ";
72 A$ = INKEY$:IF A$="" THEN 72
74 PRINT A$:IF A$ = "Y" THEN 16
80 END
```

the input to the variable A$. If no key is pressed, A$ will have nothing in it (i.e. A$ will equal "") and the instruction will be repeated. If A$ has any data in it at all, program execution passes down to line 74 where the data is printed and we check to see if A$ = "Y". If A$ equals "Y" then the execution jumps back to line 16 and we begin again; otherwise, execution passes on to line 80 which ends the program. For now this routine only provides a controlled ending of the program, but it will be used more meaningfully later.

3.2 PROGRAM EXECUTION AND PHENOMENA

In the exercises that follow we will test what we have done. The value of this section is subtle, but profound; all too often the student fails to grasp the practical significance and limitations of the subject he studies. Do you know, for example, if the results of the DFT will be exact or only approximate? Perhaps theoretically exact but masked by "noise" sources (e.g. truncation errors)? The actual results may surprise you. The following exercises have been selected to be instructive in the practical usage of the DFT. Our purpose is to gain experience of the tool we use, as well as confidence in the software we write. Our purpose is to *understand* the DFT.

3.2.1 PROGRAM EXECUTION

If we run the program created above we will obtain the results shown in Fig. 3.2 below. You will note that *only cosine components* were generated for the input function and, fortunately, only cosine components appear in the analysis; however, all of the results obtained by the analysis are one half the amplitudes of the input waveform, within the accuracy of the data printout. You will

FREQ	F(COS)	F(SIN)	Y(COS)	Y(SIN)
0	0.0000	0.0000	0.0000	0.0000
1	0.5000	0.0000	1.0000	0.0000
2	0.0000	0.0000	0.0000	0.0000
3	0.0556	0.0000	0.1111	0.0000
4	0.0000	0.0000	0.0000	0.0000
5	0.0200	0.0000	0.0400	0.0000
6	0.0000	0.0000	0.0000	0.0000
7	0.0102	0.0000	0.0204	0.0000
8	0.0000	0.0000	0.0000	0.0000
9	0.0102	0.0000	0.0000	0.0000
10	0.0000	0.0000	0.0000	0.0000
11	0.0200	0.0000	0.0000	0.0000
12	0.0000	0.0000	0.0000	0.0000
13	0.0556	0.0000	0.0000	0.0000
14	0.0000	0.0000	0.0000	0.0000
15	0.5000	0.0000	0.0000	0.0000

Figure 3.2 - Fourier Transform Output

also note that, while only the first seven harmonics were created for the input function, components show up for all 15 frequencies of the analysis. Note that the components from 9 through 15 are a "mirror image" of the components from 1 through 7 (i.e. the two halves of the spectrum are symmetrical about the Nyquest frequency). The frequencies above the Nyquest are *negative frequencies*, and consequently are complex conjugates of the frequencies below the Nyquest (as shall be seen shortly).

3.3 DATA EXERCISES

The "triangle wave" used above is a relatively simple function, but it confirms that our DFT is working. We will give our DFT program a more complicated example shortly, but before we do that, let's consider another simple test. We will analyze a single sinusoid which has been shifted in phase by 67.5° from the reference cosine wave. To do this we change the GENERATE FUNCTION Y(X) subroutine as follows:

```
122 K4=3*PI/8:KC(1)=COS(K4):KS(1)=SIN(K4):REM SET K4=67.5°
124 FOR I = 0 TO 15:K3=I*K1
126 Y(I) = COS(K3+K4)
128 NEXT I
```

At line 122 we define K4 (i.e. we set K4 to 67.5° in radians), and place the cosine and sine of this angle into KC(1) and KS(1), which is the data we will use for comparison to the output. Lines 124 through 128 then generate a full cycle of a cosine wave shifted by 67.5°. When we perform this analysis we find that the DFT yields only sine and cosine components at the fundamental and its negative. This example simply illustrates that the program can extract the sine and cosine components of a waveform that has been generated as a single sinusoid.

In most of the practical applications of the DFT we will deal with considerably more complicated functions than those presented above. A more difficult test for our program would be to create a composite time domain wave composed of completely random harmonic components—if the program can analyze this wave successfully if can handle anything. To generate this test we take advantage of the computer's ability to generate pseudo random numbers and create a random pattern of sine and cosine amplitudes. We save these amplitudes in the arrays KC(I) and KS(I), and then use them to generate the time based function Y(X). This is accomplished by changing the GENERATE FUNCTION subroutine as follows:

```
122 FOR I=0 TO 8:KC(I)=RND(1):KS(I)=RND(1):NEXT
124 FOR I=0 TO 15:FOR J=0 TO 8:K4=I*J*K1
126 Y(I)=Y(I)+KC(J)*COS(K4)+KS(J)*SIN(K4)
128 NEXT J:NEXT I
130 RETURN
```

Line 122 generates the random amplitudes of the components using the RND(1) instruction. Lines 124 through 128 then create the points of Y(I) by summing in the contributions of the sinusoids which have those random amplitudes. For each data point in the time domain function (indicated by I) we step through the 0 through 8th harmonic component contribution (indicated by J).

Now when we run the program we obtain the following
results:

FREQ	F(COS)	F(SIN)	Y(COS)	Y(SIN)
0	0.12135	0.00000	0.12135	0.65186
1	0.43443	0.36488	0.86886	0.72976
2	0.39943	0.03685	0.79885	0.07369
3	0.24516	0.22726	0.49031	0.45451
4	0.05362	0.47526	0.10724	0.95051
5	0.35193	0.26593	0.70387	0.53186
6	0.48558	0.16407	0.97116	0.32093
7	0.47806	0.46726	0.95612	0.93451
8	0.53494	0.00000	0.53493	0.56442
9	0.47806	-0.46726	0.00000	0.00000
10	0.48558	-0.16407	0.00000	0.00000
11	0.35193	-0.26593	0.00000	0.00000
12	0.05362	-0.47526	0.00000	0.00000
13	0.24516	-0.22726	0.00000	0.00000
14	0.39943	-0.03685	0.00000	0.00000
15	0.43443	-0.36488	0.00000	0.00000

Figure 3.3 - Random Amplitude Function

We notice several things about this analysis immediately:
1. The *sine* components for the 0th term and 8th term are
both zero, even though they were not zero in the input function.
This is because no sine term can exist for either of these components! SIN(N*0) = 0 and SIN(N*PI) = 0. Since we have multiplied through the zeroth term by the *frequency* of zero, all of the
sine terms will be zero in the analysis—they will also be zero in
the input function for the same reason. Likewise, there can be no
sine term for the Nyquest frequency. Even though we assigned
values to these components in our generation of the wave, they
were never created in the time domain function simply because
such components *cannot* be created.
2. The *cosine* amplitude of the zeroth and 8th frequency
components are *not* half of the input function amplitudes. Now,

we showed in the last chapter that the derived amplitudes would all be half of the actual component amplitudes, so what is going on here? The 0th term represents the D.C. component (average value) as explained in chapter 1, and all of the terms are simply multiplied by the $\cos(0) = 1$. This is reasonably apparent and, in fact, was what we should have expected; but, it may not have been apparent that the argument for the cosine term at the Nyquest frequency would always be 0 or N*PI, always yielding a cosine value of ± 1. Any cosine component in the input will be rectified, yielding an average value in the analysis equal to the peak value of that component.

3. Note that the sine components for all of the frequencies above the Nyquest are negative. This negation of the sine component comes about because the frequencies above the Nyquest are mathematically *negative* frequencies (as we noted earlier), and a negative frequency produces the complex conjugate of its positive frequency counterpart (i.e. the sine component of a complex frequency is negated but the cosine component remains unchanged).

If you are already familiar with Fourier Analysis the above observations should come as no surprise; still, it is interesting to see that practical results agree with the theory.

Let's change the GENERATE FUNCTION subroutine to illustrate one last important point: we will use a linear equation to generate a "perfect" triangle wave. We already know that the terms attenuate as $1/N^2$ in a triangle wave and that only the odd numbered terms are present—we have just analyzed a seven component approximation of this function. It would seem reasonable that we obtain similar results with a "straight line" version of this function. In the routine shown below we use a "scale factor" of $PI^2/8$ to provide the same amplitudes as the components used in our synthesized version (i.e. the fundamental has an amplitude of 1.0 and the harmonics all "roll off" as $1/N^2$). The final GENERATE FUNCTION subroutine will be:

```
122 K2=(PI*PI)/8:K3=K2/4
124 FOR I=0 TO 7:Y(I)=K2-K3*I:NEXT I
126 FOR I=8 TO 15:Y(I)=K3*I-3*K2:NEXT I
128 RETURN
```

We run the program and obtain the results shown in Fig. 3.4. First of all we notice that there are no harmonic amplitudes given for the input function; there shouldn't be any, of course, because we didn't generate the function that way. Of considerably

FREQ	F(COS)	F(SIN)	Y(COS)	Y(SIN)
0	0.00000	0.0000	0.0000	0.00000
1	0.50648	0.0000	0.0000	0.00000
2	0.00000	0.0000	0.0000	0.00000
3	0.06245	0.0000	0.0000	0.00000
4	0.00000	0.0000	0.0000	0.00000
5	0.02788	0.0000	0.0000	0.00000
6	0.00000	0.0000	0.0000	0.00000
7	0.02004	0.0000	0.0000	0.00000
8	0.00000	0.0000	0.0000	0.00000
9	0.02004	0.0000	0.0000	0.00000
10	0.00000	0.0000	0.0000	0.00000
11	0.02788	0.0000	0.0000	0.00000
12	0.00000	0.0000	0.0000	0.00000
13	0.06245	0.0000	0.0000	0.00000
14	0.00000	0.0000	0.0000	0.00000
15	0.50648	0.0000	0.0000	0.00000

Figure 3.4 - Analysis of a "Perfect" Triangle Wave

more importance is the fact that the components don't match the amplitudes that we said they would (compare them to the values derived back in Figure 3.2). Not only do they not have the correct values, they don't even have the correct $1/N^2$ ratios! This is completely wrong! Is something wrong with our program?

Even though it is customary to throw up the hands when results such as these are obtained (not a completely infrequent occurrence), and proclaim the DFT useless for "real work," it will actually be best if we can remain calm for a few minutes and examine what has happened here. First of all, there is nothing wrong with the computer program. The program is telling us exactly what it should be telling us. There is nothing wrong with the equations we used to generate the function and there is nothing wrong with the DFT we are using. The thing that is "wrong" is that we have just experienced the effects of *aliasing*! We generated the above triangle wave deliberately so that it would include the higher order harmonic components (even though we know the maximum Nyquest frequency is "8" for a data base of only 16 data points. All of the harmonics above that frequency have been "folded back" into the spectrum we are analyzing and have given us "incorrect" values. If we want to generate a function as we did in this example, and have it agree with the known harmonic analysis of the "classic" waveshape, we must *filter off* the harmonics above the nyquest before we attempt to *digitize* it.

The mechanics of the aliasing phenomenon are very interesting to delve into, but our concern here is with the FFT, and so we will resist the urge to dig deeper. We have gone through this exercise because it is the sort of thing that happens in the practical application of DFT/FFT routines. Systems are improperly designed (or improperly applied) and then, later, no one can understand why the results are invalid. The DFT algorithm is indeed a simple program, but there are a great many "traps" lurking for the unwary. While it is relatively easy to explain how the FFT works and how to write FFT programs, there is no alternative to studying the DFT, and FFT, and all of the associated engineering disciplines, in detail. Like Geometry (and, for that matter, most other things of value)—for this subject "There is no royal road ..."

CHAPTER IV

THE INVERSE TRANSFORM
AND COMPLEX VARIABLES

4.1 RECONSTRUCTION

The inverse transform is, intuitively, a very simple operation. We know what the amplitudes of the sinusoids are (from the forward transform), so we simply reconstruct all of these sinusoids and sum them together. Nothing could be simpler.

We note that the process of extracting the individual frequency components yielded only half amplitude values for all but the constant term and the Nyquest frequency term; however, we extracted components for both the negative and positive frequencies (i.e. both above and below the Nyquest). This all works out neatly in the reconstruction process since it will provide precisely the correct amplitudes when both negative and positive frequency terms are summed in. Before we develop this discussion further let's write an inverse transform routine and incorporate it into the DFT program of the preceding chapters.

```
6 REM    **********************************************
8 REM    ** (DFT4.1) ANALYZE/RECONSTRUCT WAVEFORM **
10 REM   **********************************************
11 REM   *** DEFINE CONSTANTS
12 PI=3.141592653589793#:P2=2*PI:K1=PI/8:K2=1/PI
13 REM   *** DIMENSION ARRAYS
14 DIM Y(16),FC(16),FS(16),KC(16),KS(16),Z(16)
15 REM   *** INITIALIZE FOURIER COEFFICIENT ARRAYS
16 CLS:FOR J=0 TO 16:FC(J)=0:FS(J)=0:NEXT
20 GOSUB 108: REM * PRINT COLUMN HEADINGS
30 GOSUB 120: REM * GENERATE FUNCTION Y(X)
40 GOSUB 200: REM * PERFORM DFT
60 GOSUB 140: REM * PRINT OUT FINAL VALUES
69 REM   *** ASK IF RECONSTRUCTION IS NECESSARY
70 PRINT:PRINT "RECONSTRUCT (Y/N)? ";
72 A$ = INKEY$:IF A$="" THEN 72
74 PRINT A$:IF A$ = "Y" THEN 80
76 END
80 CLS:GOSUB 220:REM * RECONSTRUCT
82 GOSUB 240:REM * PRINT OUTPUT
84 PRINT:PRINT "MORE (Y/N)?";
86 A$ = INKEY$:IF A$ = "" THEN 86
88 PRINT A$:IF A$ = "Y" THEN 15
90 GOTO 76
100 REM  ******************************************
102 REM  *         PROGRAM SUBROUTINES           *
104 REM  ******************************************
106 REM  *      PRINT COLUMN HEADINGS            *
108 PRINT:PRINT
109 REM  *** Y(COS) AND Y(SIN)=INPUT COMPONENT AMPLITUDES
110 PRINT "FREQ   F(COS)     F(SIN)     Y(COS)     Y(SIN)"
112 PRINT
114 RETURN
118 REM  ******************************????
120 REM  *** GENERATE FUNCTION F(X) ***
122 FOR I = 0 TO 15:K3=I*K1:REM I=DATA POINT LOCATION IN ARRAY
123 REM  *** SET Y(I)=FIRST 8 COMPONENTS OF TRIANGLE WAVE
124 Y(I) = COS(K3)+COS(3*K3)/(9)+COS(5*K3)/(25)+COS(7*K3)/49
126 NEXT
127 REM  *** STORE COMPONENT AMPLITUDES
128 FOR I=1 TO 7 STEP 2: KC(I)=1/I^2:NEXT
130 RETURN
```

```
132 REM ******************************
138 REM *       PRINT OUTPUT        *
140 FOR Z=0 TO 15
142 PRINT Z;"   ";:REM * Z=COMPONENT FREQUENCY
144 PRINT USING "##.#####_   ";FC(Z),FS(Z),KC(Z),KS(Z)
146 NEXT Z
148 RETURN
200 REM ***************************
202 REM * SOLVE FOR COMPONENTS *
206 FOR J=0 TO 15:REM * SOLVE EQNS FOR EACH FREQUENCY
208 FOR I = 0 TO 15:REM * MULTIPLY AND SUM EACH DATA POINT
210 FC(J)=FC(J)+Y(I)*COS(J*I*K1):FS(J)=FS(J)+Y(I)*SIN(J*I*K1)
212 NEXT I
214 FC(J)=FC(J)/16: FS(J)=FS(J)/16:REM * FIND MEAN VALUE
216 NEXT J
218 RETURN
220 REM **************************
222 REM *      RECONSTRUCT      *
224 REM **************************
226 FOR J=0 TO 15:REM * RECONSTRUCT EACH FREQUENCY
228 FOR I = 0 TO 15: REM * RECONSTRUCT EACH DATA POINT
230 Z(I)=Z(I)+FC(J)*COS(J*I*K1)+FS(J)*SIN(J*I*K1)
232 NEXT I
234 NEXT J
236 RETURN
240 REM ******************************
241 REM *       PRINT OUTPUT        *
240 REM ******************************
243 REM * Y(I) EQUALS INPUT FUNCTION FOR COMPARISON
244 CLS:PRINT:PRINT "T       Z(I)       Y(I)":PRINT:PRINT
245 FOR Z=0 TO 15
246 PRINT Z;"   ";
248 PRINT USING "##.#####_   ";Z(Z),Y(Z)
250 NEXT Z
252 RETURN
```

Figure 4.1

The first part of this program is apparently unchanged from the program of the preceding chapter, except that line 14 defines a new array Z(16). This array will hold the reconstructed input function. At line 70 we change the question asked to the following: "RECONSTRUCT (Y/N)". If the answer is "Y" then we pass on to line 80, where we begin the operation of reconstruction.

As in the preceding program, we use subroutines to simplify the operation. At line 80 we jump down to line 220 where the operation of reconstruction is performed. At line 82 we print out the results. Line 84 asks if we want "MORE ?". A "Y" returns us to line 16—anything else ends the program. Let's now look at the inverse transform routine:

```
226 FOR J=0 TO 15:REM * RECONSTRUCT EACH FREQUENCY
228 FOR I = 0 TO 15: REM * RECONSTRUCT EACH DATA POINT
230 Z(I)=Z(I)+FC(J)*COS(J*I*K1)+FS(J)*SIN(J*I*K1)
232 NEXT I
234 NEXT J
236 RETURN
```

At line 226 we set up a loop to count from 0 to 15 (i.e. count the frequency components used in the reconstruction). At line 228 we set up a nested loop to count through the 16 data points of the reconstruction. At line 230 we sum into the array Z(I) the contribution of the Jth frequency component at the data point I (both cosine and sine components). As pointed out above we sum in *all* of the frequency components, i.e. the contribution from both the positive and negative frequency components, the constant term, and the Nyquest frequency term. It's that simple.

The print routine for the reconstructed function Z(Z) (as well as the input time domain function Y(Z)) is located at line 240.

```
243 REM * Y(I) EQUALS INPUT FUNCTION FOR COMPARISON
244 CLS:PRINT:PRINT "T        Z(I)        Y(I)":PRINT:PRINT
245 FOR Z=0 TO 15
246 PRINT Z;"    ";
248 PRINT USING "##.#####_    ";Z(Z),Y(Z)
250 NEXT Z
```

At line 244 we clear the screen and print the new heading. At lines 245 through 250 we print the reconstruction as well as the input data.

If we run this program we will obtain the following output:

T	Z(I)	Y(I)
0	1.17152	1.17152
1	0.93224	0.93224
2	0.61469	0.61469
3	0.30918	0.30918
4	-0.00000	-0.00000
5	-0.30918	-0.30918
6	-0.61469	-0.61469
7	-0.93224	-0.93224
8	-1.17152	-1.17152
9	-0.93224	-0.93224
10	-0.61469	-0.61469
11	-0.30918	-0.30918
12	-0.00000	-0.00000
13	0.30918	0.30918
14	0.61468	0.61469
15	0.93224	0.93224

Figure 4.2

4. 2 TRANSFORM SYMMETRY AND COMPLEX VARIABLES

All of the above is beautifully simple; unfortunately, the purist will never let us leave things that way. While the above is certainly *not incorrect*, it is slightly "out of bed" with the formal definition of the DFT. Actually, the complications are not all that difficult; we need only reformulate everything in terms of complex variables. Let's look at the problem:

The definitions of the DFT and Inverse DFT are:

$$F(f) = 1/N \sum_{T=0}^{N-1} f(t)\, W_N^{-fT} \quad \text{------------------------} \quad (4.1)$$

$$f(T) = \sum_{f=0}^{N-1} F(f)\, W_N^{Tf} \quad \text{---------------------------} \quad (4.2)$$

Where: $F(f)$ = frequency components or transform
              $f(T)$ = time base data points or inverse xform
              N = number of data points
              T = discrete times
              f = discrete frequencies
              $W_N = e^{i2\pi/N} = Cos(2\pi/N) + i\, Sin(2\pi/N)$

There is marked symmetry between eqns. (4.1) and (4.2), but the algorithms given in DFT4.1 for the transform and inverse transform fail to reflect that symmetry. The *inverse transform* starts from complex quantities in the frequency domain while we use only real numbers for the input function in the *forward transform*. Now, in the general case, both the frequency domain and the time domain may be complex numbers, of course, and when we provide for this potential, the symmetry between the forward and inverse transforms immediately becomes apparent. Let's look at what happens to our transform algorithm when we consider complex variables:

Multiplication of two complex quantities yields the following terms:

$$(A +iB)(C +iD) = AC +iAD +iBC - BD$$
$$= (AC-BD)+i(AD+BC) \text{----------} \quad (4.3)$$

We might note that $(A +iB)$ is the input function, and $(C +iD)$ is equal to $W_N = e^{i2\pi/N} = Cos(2\pi/N) + i\,Sin(2\pi/N)$. Incorporating this into program DFT4.1, we convert the forward transform algorithm:

```
210 FC(J) = FC(J)+YR(I)*COS(J*I*K1)-YI(I)*SIN(J*I*K1)
211 FS(J) = FS(J)+YR(I)*SIN(J*I*K1)+YI(I)*COS(J*I*K1)
```

where YR stands for the real part of the input function and YI stands for the imaginary part (this obviously requires defining new arrays, YR(16) and YI(16), at program initialization). Similarly, recognizing that an imaginary term must be created in the inverse transform (as defined in eqn. 4.2), the reconstruction algorithm becomes:

```
230 ZR(I) = ZR(I)+FC(J)*COS(J*I*K1)+FS(J)*SIN(J*I*K1)
231 ZI(I) = ZI(I)-FC(J)*SIN(J*I*K1)+FS(J)*COS(J*I*K1)
```

The symmetry is now much more apparent. Except for the sign changes, these lines of code are identical. With a little manipulation we can use the same routine for both forward and inverse transformation.

Very well then, we will write a new DFT program with the above considerations incorporated. It will require completely revamping the data structures and even the basic flow of the program, but it will be formally correct. For us, at this point, it illustrates a significant characteristic of the DFT/Inverse DFT. While we are at it we should change the program to a menu driven format as it will be much better suited to our work in the following chapters. The new program (DFT4.2) is shown on the following pages. We must make note of the changes:

1. The data arrays have changed completely. At line 14 we now dimension four arrays—C(2,16), S(2,16), KC(2,16), and

```
14 DIM C(2,16),S(2,16),KC(2,16),KS(2,16)
```

KS(2,16). These are "two dimensional" arrays, if you will. There are two columns of 16 data points in each array. From now on we will put the time domain data in column 1 and the frequency domain data in column 2 of each of these arrays. KC(2,16) and

```
6 REM    *******************************************
8 REM    *** (DFT4.2) GENERATE/ANALYZE WAVEFORM ***
10 REM   *******************************************
12 PI=3.141592653589793#:P2=2*PI:K1=PI/8:K2=1/PI
14 DIM C(2,16),S(2,16),KC(2,16),KS(2,16)
16 CLS:FOR J=0 TO 16:FOR I=1 TO 2:C(I,J)=0:S(I,J)=0:NEXT:NEXT

19 REM      *******************
20 CLS:REM *    MAIN MENU     *
21 REM      *******************
22 PRINT:PRINT:PRINT "        MAIN MENU":PRINT
24 PRINT " 1 = GENERATE FUNCTION":PRINT
26 PRINT " 2 = TRANSFORM FUNCTION":PRINT
28 PRINT " 3 = INVERSE TRANSFORM":PRINT
30 PRINT " 4 = EXIT":PRINT:PRINT
32 PRINT SPC(10);"MAKE SELECTION";
34 A$ = INKEY$:IF A$="" THEN 34
36 A=VAL(A$):ON A GOSUB 300,40,80,1000
38 GOTO 20

39 REM ******************************
40 REM * FORWARD TRANSFORM ROUTINE *
41 REM ******************************
42 CLS:N=1:M=2:K5=16:K6=-1:GOSUB 108
44 FOR J=0 TO 16:C(2,J)=0:S(2,J)=0:NEXT
45 GOSUB 200: REM - PERFORM DFT
46 GOSUB 140: REM - PRINT OUT FINAL VALUES
48 PRINT:INPUT "C/R TO CONTINUE";A$
50 RETURN

79 REM **************************
80 REM *   INVERSE TRANSFORM    *
81 REM **************************
82 CLS:FOR I=0 TO 15:C(1,I)=0:S(1,I)=0:NEXT
84 N=2:M=1:K5=1:K6=1:GOSUB 200:REM RECONSTRUCT INPUT
85 GOSUB 150:REM PRINT HEADING
86 GOSUB 140:REM PRINT OUTPUT
88 PRINT:INPUT "C/R TO CONTINUE";A$
90 RETURN
```

```
100 REM  ****************************************
102 REM  *        PROGRAM SUBROUTINES        *
104 REM  ****************************************
106 REM  *      PRINT COLUMN HEADINGS        *
108 PRINT:PRINT
110 PRINT "FREQ    F(COS)    F(SIN)    Y(COS)    Y(SIN)"
112 PRINT
114 RETURN

137 REM  ******************************
138 REM  *      PRINT OUTPUT        *
139 REM  ******************************
140 FOR Z=0 TO 15
142 PRINT Z;"   ";
144 PRINT USING "##.#####_   ";C(M,Z),S(M,Z),KC(M,Z),KS(M,Z)
146 NEXT Z
148 RETURN

150 REM  ******************************
152 REM  *   PRINT COLUMN HEADINGS   *
154 PRINT:PRINT
156 PRINT " T        RECONSTRUCTION        INPUT FUNCTION"
158 PRINT
160 RETURN

200 REM  ******************************
202 REM  *    TRANSFORM/RECONSTRUCT   *
204 REM  ******************************
206 FOR J=0 TO 15:REM SOLVE EQNS FOR EACH FREQUENCY
208 FOR I=0 TO 15:REM MULTIPLY AND SUM EACH POINT
210 C(M,J)=C(M,J)+C(N,I)*COS(J*I*K1)+K6*S(N,I)*SIN(J*I*K1)
211 S(M,J)=S(M,J)-K6*C(N,I)*SIN(J*I*K1)+S(N,I)*COS(J*I*K1)
212 NEXT I
214 C(M,J)=C(M,J)/K5:S(M,J)=S(M,J)/K5:REM SCALE RESULTS
216 NEXT J
218 RETURN

299 REM     ***********************
300 CLS:REM *   FUNCTION MENU    *
301 REM     ***********************
302 FOR I=0 TO 15:C(1,I)=0:S(1,I)=0
303 FOR J=1 TO 2:KC(J,I)=0:KS(J,I)=0:NEXT:NEXT
```

```
304 PRINT:PRINT:PRINT "        FUNCTION MENU":PRINT
306 PRINT " 1 = TRIANGLE WAVE":PRINT
308 PRINT " 2 = CIRCLE":PRINT
310 PRINT " 3 = ELLIPSE 1":PRINT
312 PRINT " 4 = ELLIPSE 2":PRINT:PRINT
320 PRINT SPC(10);"MAKE SELECTION";
322 A$ = INKEY$:IF A$="" THEN 322
326 A=VAL(A$):ON A GOSUB 330,340,350,360,1000
328 RETURN


330 REM *** GENERATE FUNCTION F(X) ***
332 FOR I = 0 TO 15:K3=I*K1
334 C(1,I) = COS(K3)+COS(3*K3)/9+COS(5*K3)/25+COS(7*K3)/49
335 KC(1,I)=C(1,I)
336 NEXT
338 FOR I=1 TO 7 STEP 2:KC(2,I)=1/I^2:NEXT
339 RETURN
340 REM *** GENERATE CIRCLE ***
342 FOR I = 0 TO 15:K3=I*K1
344 C(1,I) = SIN(K3):S(1,I)=COS(K3)
345 KC(1,I)=C(1,I):KS(1,I)=S(1,I)
346 NEXT
348 KS(2,1)=1
349 RETURN
350 REM *** GENERATE ELLIPSE 1 ***
352 FOR I = 0 TO 15:K3=I*K1
354 C(1,I) = SIN(K3):S(1,I)=2*COS(K3)
355 KC(1,I)=C(1,I):KS(1,I)=S(1,I)
356 NEXT
358 KS(2,1)=1.5:KS(2,15)=.5
359 RETURN
360 REM *** GENERATE ELLIPSE 2 ***
362 FOR I = 0 TO 15:K3=I*K1
364 C(1,I) = COS(K3):S(1,I)=2*SIN(K3)
365 KC(1,I)=C(1,I):KS(1,I)=S(1,I)
366 NEXT
368 KC(2,1)=-.5:KC(2,15)=1.5
369 RETURN
1000 STOP
```

Figure 4.3

KS(2,16) are not needed for a "working" program, but we use
them here to save the input functions which we have generated.
We use these later for comparison with the transform and inverse
transform. Again, the first column stores the time domain data and
the second stores the frequency domain.

2. The program is menu driven. Lines 20 through 30 print
the menu. Lines 32 and 34 determine what the selection is and

```
20 CLS:REM *    MAIN MENU    *
21 REM       ********************
22 PRINT:PRINT:PRINT "        MAIN MENU":PRINT
24 PRINT " 1 = GENERATE FUNCTION":PRINT
26 PRINT " 2 = TRANSFORM FUNCTION":PRINT
28 PRINT " 3 = INVERSE TRANSFORM":PRINT
30 PRINT " 4 = EXIT":PRINT:PRINT
32 PRINT SPC(10);"MAKE SELECTION";
34 A$ = INKEY$:IF A$="" THEN 34
36 A=VAL(A$):ON A GOSUB 300,40,80,1000
38 GOTO 20
```

jump to the appropriate subroutine. The "generate function" is still
located at line 120. The transform and inverse transform routines
are now located at lines 40 and 80 respectively.

3. The "transform routine" is now located at line 40.
Since we now store the time and frequency data in the same array,
and since the forward and inverse transforms are performed by the
same sub-routine, we must set up "pointers" so that the transform
routine will know which way to operate on the data. We do this by
using M and N as the pointers. N points to the "input function"
and M points to the output function—if N=1 and M=2 (according
to the statement above that 1 was time domain and 2 was frequency domain data) then we will perform a "Forward Transform." If

N=2 and M=1 we will perform the "Inverse Transform." In either case, we must have already created the input function (i.e. we must use the generate function option from the Main Menu) before we can perform a forward transform, and we must have performed a forward transform before we perform an inverse transform. From these conditions it is apparent what we must do to perform a forward transform: we clear the screen at line 40, set N=1, M=2, and then jump to line 108 to print the heading for the output. You will also note that we have set the constant K5=16. This constant is used to find the average value of each transformed component as we did in line 214 of DFT4.1 (Fig. 4.1). Since we do not need to make this division in the inverse transform, we set K5=1 for that operation (see inverse transform of Fig. 4.3 above).

At line 42 we clear the frequency domain arrays (i.e. $C(2,16)$ and $S(2,16)$) before performing the transform. At line 44 we then jump down to line 200 where we perform the DFT on the time domain data. After performing the DFT we return to line 46, where we then jump to line 140 and print the results. Line 48 is only a programming technique for waiting until the user is through examining the data before returning to the main menu.

4. The "inverse transform" starts at line 80 and follows the same pattern as the forward transform.

5. As we noted above, the transform routine starts at line 200. It is similar to the transform routines used previously except now we include the possibility of transforming complex numbers.

4.3 PROGRAM OPERATION/EXAMPLES

If we run this program for the familiar triangle wave of our past examples we will obtain the same data that we obtained previously (Fig 4.2) except that now zeros will be printed in the column for the imaginary part of the input function (i.e. we still create only the real part of the time domain function).

While this works, of course, we might want to check this program for some function which actually has complex numbers for the input. The GENERATE FUNCTION subroutine presents a second menu which offers a selection of functions. We may take sixteen points on the circumference of a circle as an example—or perhaps the example of an ellipse would be more interesting. If you have not worked with the Fourier Transform of complex variable inputs the results of these examples might prove interesting.

We are *not* primarily concerned with the transform of complex variables in this short book, and so we will complete our review of the DFT here. It is interesting to note (in connection with complex variables) that we may actually take the transform of two *real valued functions* simultaneously by placing one in the real part of the input array and the other in the imaginary part. By relatively simple manipulation of the output each of the individual spectrums may be extracted (see, for example, *Fast Fourier Transforms*, Chapter 3.5, by J.S. Walker, *CRC PRESS*).

PART II

THE FFT

CHAPTER V

FOUR FUNDAMENTAL THEOREMS

5.0 INTRODUCTION

Our development of the FFT (in chapter 7) will be based on these four theorems. Their *validity* is not our concern here (proofs are relegated to appendix 5.3); rather, we need only understand their function. We will illustrate these theorems via real examples using the DFT program developed in the previous chapters. This DFT program has necessarily been expanded for these illustrations and is listed in appendices 5.1 and 5.2.

This material is easily grasped, but that does not diminish its importance—its comprehension is imperative. These theorems are the key to understanding the FFT, and consequently, this chapter is dedicated solely to walking through each of these illustrations step by step. The best approach might be to run each illustration on your computer while reading the accompanying text.

5.1 THE SIMILARITY THEOREM

The Similarity Theorem might better be called "the reciprocity theorem" for it states: "As the time domain function expands in time, the frequency domain function compresses in spectrum, and increases in amplitude." The input function for program DFT5.01 is a half cycle of $\sin^2(x)$ centered in the middle of the time domain. The program requests a "width" (actually the half-width) which specifies the number of data points over which the input function is to be spread. According to Similarity then, the *spectrum* of the frequency domain will expand and compress in *inverse proportion* to the specified width of the time domain function. The *amplitude* of the time domain function is held constant (peak amplitude of 32) so that, in keeping with Similarity,

T			T		
0	+0.00000	+0.00000	16	+32.00000	+0.00000
1	+0.00000	+0.00000	17	+0.00000	+0.00000
2	+0.00000	+0.00000	18	+0.00000	+0.00000
3	+0.00000	+0.00000	19	+0.00000	+0.00000
4	+0.00000	+0.00000	20	+0.00000	+0.00000
5	+0.00000	+0.00000	21	+0.00000	+0.00000
6	+0.00000	+0.00000	22	+0.00000	+0.00000
7	+0.00000	+0.00000	23	+0.00000	+0.00000
8	+0.00000	+0.00000	24	+0.00000	+0.00000
9	+0.00000	+0.00000	25	+0.00000	+0.00000
10	+0.00000	+0.00000	26	+0.00000	+0.00000
11	+0.00000	+0.00000	27	+0.00000	+0.00000
12	+0.00000	+0.00000	28	+0.00000	+0.00000
13	+0.00000	+0.00000	29	+0.00000	+0.00000
14	+0.00000	+0.00000	30	+0.00000	+0.00000
15	+0.00000	+0.00000	31	+0.00000	+0.00000

Fig. 5.1 - Time Domain for Width = 1

the spectrum amplitude will be *proportional* to the width! In this example the amplitude of the output will vary between 1.0 and 16 (for the range of widths allowed—also 1 through 16). Repeating the example with various widths illustrates Similarity.

Run DFT5.01 and specify a width of 1. A single data point will be generated as the input function (we reproduce the computer screen in fig. 5.1 on the previous page). The spectrum for this input is "flat" (i.e. a series of components alternating between +1 and -1 as shown in figure 5.2 below). A graph of both frequency and time domain functions is given in figure 5.3. Note that in the graphical display only the *magnitude* of the frequency domain data is displayed.

FREQ	F(COS)	F(SIN)	FREQ	F(COS)	F(SIN)
0	+1.00000	+0.00000	16	+1.00000	+0.00000
1	-1.00000	-0.00000	17	-1.00000	-0.00000
2	+1.00000	+0.00000	18	+1.00000	+0.00000
3	-1.00000	-0.00000	19	-1.00000	-0.00000
4	+1.00000	+0.00000	20	+1.00000	+0.00000
5	-1.00000	-0.00000	21	-1.00000	-0.00000
6	+1.00000	+0.00000	22	+1.00000	+0.00000
7	-1.00000	-0.00000	23	-1.00000	-0.00000
8	+1.00000	+0.00000	24	+1.00000	+0.00000
9	-1.00000	-0.00000	25	-1.00000	-0.00000
10	+1.00000	+0.00000	26	+1.00000	+0.00000
11	-1.00000	-0.00000	27	-1.00000	-0.00000
12	+1.00000	+0.00000	28	+1.00000	+0.00000
13	-1.00000	-0.00000	29	-1.00000	-0.00000
14	+1.00000	+0.00000	30	+1.00000	+0.00000
15	-1.00000	-0.00000	31	-1.00000	-0.00000

Fig 5.2 - Spectrum for Time Domain Width = 1

This single example doesn't say anything about the Similarity Theorem of course for the relationship concerns the expansion and compression of the function. Repeat the exercise

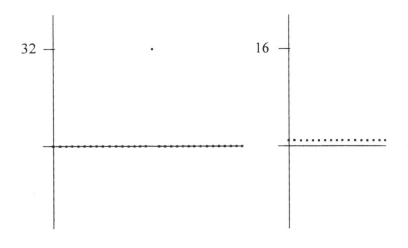

Time Domain Waveshape   Frequency Domain

Fig. 5.3 - Similarity Test Width = 1

but this time select a width of 2. The graphical results are shown
in Fig. 5.4 below. This display shows the "expanded" time domain
function which now has a maximum value at the 16th data point
(amplitude = 32) with two additional data points (amplitudes = 16)
on either side. The spectrum is still a series of data points which
alternate in sign, but now the amplitude of the frequency components diminish as the frequency increases. At a frequency of 8 the
amplitude is 0.5, and from there the components "roll off" to
negligible amplitudes at the higher frequencies.

     Continue the experiment by repeating the Similarity Test
with widths of 4, 8 and 16. The frequency spectrum shrinks and
the amplitude increases as the time domain function expands (see
figures 5.5 through 5.7).

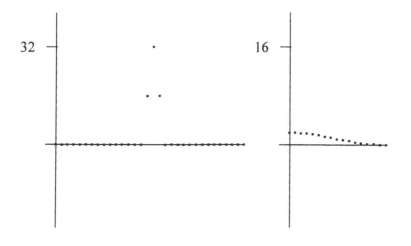

Time Domain Waveshape   Frequency Domain

Fig. 5.4 - Similarity Test Width = 2

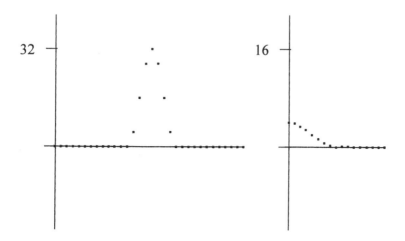

Time Domain Waveshape   Frequency Domain

Fig. 5.5 - Similarity Test Width = 4

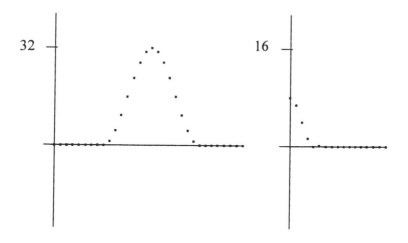

Time Domain Waveshape   Frequency Domain

Fig 5.6 - Similarity Test Width = 8

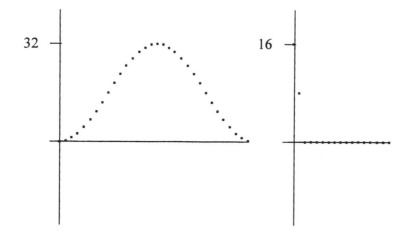

Time Domain Waveshape   Frequency Domain

Fig. 5.7 - Similarity Test Width = 16

This phenomenon is the relationship known as Similarity. It is understood, of course, that "similarity" is completely general (i.e. it works for any input function) and also bilateral; if we compress the *spectrum* of a function its *time domain* will be expanded and simultaneously decreased in amplitude. This relationship is indeed simple, but not insignificant nor trivial. In fact, it is perhaps the most fundamental relationship that exists between the frequency and time domains. The essence of this relationship is this: Faster transitions and shorter durations require (imply) higher frequencies, and slower transitions and longer durations require (imply) lower frequencies. If your tape recorder runs too fast, everyone sounds like Chip and Dale; if it runs too slow they sound like Lurch. It's a relationship that's inevitable—still, perhaps, not *completely* inescapable.

5.2 THE ADDITION THEOREM

This theorem states that the transform of the sum of two functions is equal to the sum of the transforms of the two functions individually:

$$\text{Xform } \{f1(x)+f2(x)\} = \text{Xform } \{f1(x)\} + \text{Xform } \{f2(x)\} \text{ ----} \quad (5.1)$$

This is the result of the system being linear of course, and consequently, may not seem remarkable. On the other hand, it allows a certain amount of manipulation that is worth illustrating.

The example selected for DFT5.02 concerns "rising" and

"falling" exponential functions. The rising edge is described by:

$$f_1(x) = A_0 (1 - e^{-x/T}) \quad \text{-------------------------} \quad (5.2)$$
$$A_0 = \text{Amplitude of final value}$$
$$T = \text{Time Constant}$$

But the falling edge is described by:

$$f_2(x) = A_1 \ e^{-x/T} \quad \text{-----------------------------} \quad (5.3)$$
$$A_1 = \text{Starting Amplitude}$$

DFT5.02 is slightly longer than the other programs of this chapter simply because there are more things to do. We first generate the *leading edge* exponential (eqn. 5.1), find the transform, and display the results (both printing out the numerical values and graphically plotting the results). Before we generate the second input function (i.e. the trailing edge exponential), we save this frequency domain data for the rising edge. We then generate and transform the *falling edge* input function (eqn. 5.2) and again print and plot the results. We are now ready to illustrate the Addition Theorem—we sum the two frequency domain functions and take the inverse transform.

The second part of the demonstration consists of summing the two time domain functions, taking the transform of this summation, and printing and plotting this result.

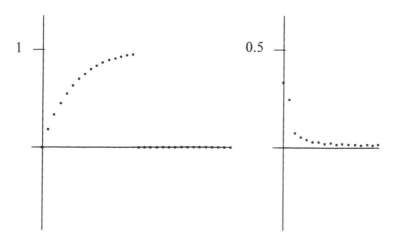

Time Domain Waveshape   Frequency Spectrum

Fig. 5.8 - $f_1(x) = A_0 (1 - e^{-x/T})$

We show the results of the "leading edge" transform as well as the time domain input function in fig. 5.8. Note that in the plot of the frequency domain (and in the following tables) we show the frequency spectrum only up to the Nyquest frequency as the components above that frequency are essentially redundant.

The function $f_2(x)$ (i.e. the "falling edge") and its transform are shown below in figure 5.9. Once again, these two functions (i.e. fl(x) and f2(x)) and their transforms tell us nothing about the theorem we are trying to illustrate; they are only two slightly interesting looking functions, displaced in time so that one begins where the other ends. For the actual illustration we add the two transforms together and take the inverse transform (fig. 5.10). The reconstruction from the frequency domain summation is shown in figure 5.11 and the equivalent time domain is shown in 5.12.

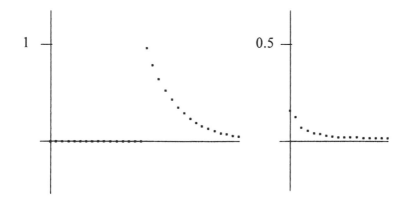

Time Domain Waveshape   Frequency Spectrum

Fig. 5.9 - $f_2(x) = A1\ e^{-x/T}$

When we add the two transforms of the separate functions
and take the inverse transform, we get the perfect combination of the
two functions in the time domain.

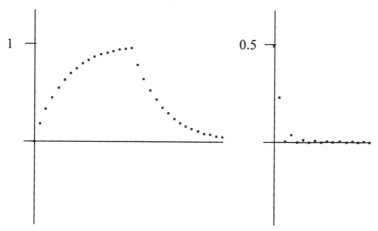

Time Domain Waveshape   Frequency Spectrum

Fig. 5.10 - $f_1(x)+f_2(x)$

T			T		
0	+0.00000	-0.00000	16	+0.95924	-0.00000
1	+0.18127	+0.00000	17	+0.78536	-0.00000
2	+0.32968	+0.00000	18	+0.64300	-0.00000
3	+0.45119	+0.00000	19	+0.52644	+0.00000
4	+0.55067	-0.00000	20	+0.43101	+0.00000
5	+0.63212	-0.00000	21	+0.35288	+0.00000
6	+0.69881	-0.00000	22	+0.28892	-0.00000
7	+0.75340	-0.00000	23	+0.23654	+0.00000
8	+0.79810	-0.00000	24	+0.19367	+0.00000
9	+0.83470	-0.00000	25	+0.15856	+0.00000
10	+0.86466	+0.00000	26	+0.12982	+0.00000
11	+0.88920	+0.00000	27	+0.10629	+0.00000
12	+0.90928	+0.00000	28	+0.08702	+0.00000
13	+0.92573	-0.00000	29	+0.07125	+0.00000
14	+0.93919	+0.00000	30	+0.05833	-0.00000
15	+0.95021	+0.00000	31	+0.04776	+0.00000

Fig. 5.11 - Reconstruction from Sum of Transforms

T			T		
0	+0.00000	-0.00000	16	+0.95924	-0.00000
1	+0.18127	+0.00000	17	+0.78536	-0.00000
2	+0.32968	+0.00000	18	+0.64300	-0.00000
3	+0.45119	+0.00000	19	+0.52644	+0.00000
4	+0.55067	-0.00000	20	+0.43101	+0.00000
5	+0.63212	-0.00000	21	+0.35288	+0.00000
6	+0.69881	-0.00000	22	+0.28892	-0.00000
7	+0.75340	-0.00000	23	+0.23655	+0.00000
8	+0.79810	-0.00000	24	+0.19367	+0.00000
9	+0.83470	-0.00000	25	+0.15856	+0.00000
10	+0.86466	+0.00000	26	+0.12982	+0.00000
11	+0.88920	+0.00000	27	+0.10629	+0.00000
12	+0.90928	+0.00000	28	+0.08702	+0.00000
13	+0.92573	-0.00000	29	+0.07125	+0.00000
14	+0.93919	+0.00000	30	+0.05833	-0.00000
15	+0.95021	+0.00000	31	+0.04776	+0.00000

Fig. 5.12 - Sum of $f_1(x)$ and $f_2(x)$

Part 2 of the illustration is essentially redundant. We add the two time based functions and take the transform. If you run this part of DFT5.02 you will find we get identical results with the previous example, which was, in fact, what the theorem proposed. The transform of the sum of two functions is equal to the sum of the transforms of the individual functions. It is apparent that this is a bilateral relationship.

5.3 THE SHIFTING THEOREM

This theorem states that if a time domain function is shifted in time, the amplitude of the frequency components will remain constant, but the phases of the components will shift linearly—proportional to both the frequency of the component and the amount of the time shift.

In program DFT5.03 we have modified the printout routine so that the magnitude and phase of the frequency components can be printed as opposed to printing out the sine and cosine components (we find the magnitude by the RSS [Root of the Sum of the Squares] of the components, and the phase as the Arc Tangent of the ratio of the two components).

This theorem uses the Impulse Function, which is a unique function ideally suited for our purpose. The Impulse Function is a pulse whose width approaches zero, and amplitude approaches infinity, while its "area" (i.e. product of width x amplitude) remains fixed. This unique function produces a unique spectrum—the frequency components all have amplitudes of 1.0 when the area of the function is unity. Now, it is obviously impossible to represent an infinite amplitude on a computer, but fortunately, in the DFT, we don't really need an infinite amplitude. If we make the

FREQ	F(MAG)	F(THETA)
0	+1.00000	+0.00000
1	+1.00000	+0.00000
2	+1.00000	+0.00000
3	+1.00000	+0.00000
4	+1.00000	+0.00000
5	+1.00000	+0.00000
6	+1.00000	+0.00000
7	+1.00000	+0.00000
8	+1.00000	+0.00000
9	+1.00000	+0.00000
10	+1.00000	+0.00000
11	+1.00000	+0.00000
12	+1.00000	+0.00000
13	+1.00000	+0.00000
14	+1.00000	+0.00000
15	+1.00000	+0.00000
16	+1.00000	+0.00000

Fig. 5.13 - Impulse Xform

amplitude of the impulse equal to the number of data points in the digitized array, we will obtain the desired results (i.e. the frequency components will all have an amplitude of 1.0- see figs. 5.13 and 5.14).

Run program DFT5.03 and select the Shifting Theorem from the MAIN MENU; the routine automatically runs the first example with a time shift of zero. Specifically, the computer will generate a time domain impulse of amplitude 32, followed by 31 data points

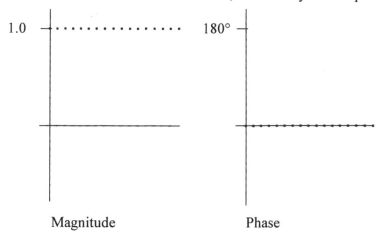

Magnitude   Phase

Fig. 5.14 - Impulse Xform (polar coordinates)

of zero amplitude. It then takes the transform and prints the output (in polar coordinates). The results are shown in fig. 5.14—all of the magnitudes will be 1.0 and the phases will be zero (note that the phase is given in degrees).

Now, a "shift" of 1 will cause the impulse to be generated at the second data point (i.e. it is "shifted" to the right by one data point). According to the Shifting Theorem, the magnitude of the frequency components will be unchanged, but their phases will be shifted proportional to the time base shift. Furthermore, the phase

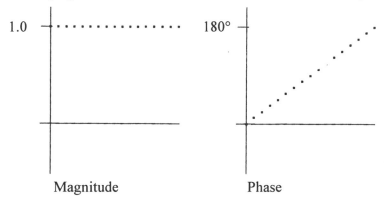

Fig. 5.15 - Xform of Shifted Impulse

of each component will be proportional to its harmonic number. The results will be as shown in Fig. 5.15. Note that the phase shift increases by 11.25° for each frequency component yielding a shift of π radians (180°) at the Nyquest frequency. This is a convenient feature of the DFT—the phase shift at the Nyquest frequency will always be π radians multiplied by the number of data points that the time domain function has been shifted. The phase continues to shift in this manner as we continue upward through the negative frequencies of course. We may verify this fact by running the Shifting Theorem routine and selecting various shifts. A time shift of two data points, for example, yields increments of 22.5° for each component up to 360 degrees at the Nyquest frequency, or by

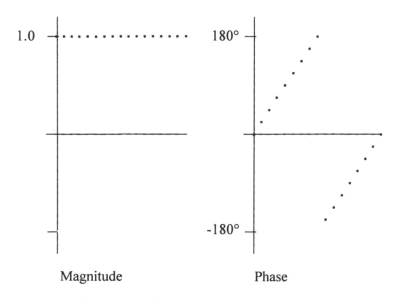

Magnitude   Phase

Fig. 5.16 - Xform of Double Shifted Impulse

twice the amount that it did for a single data point shift (fig. 5.16).

The principle is (again) completely general and bilateral. You might want to modify the routine to insert some other function for evaluation. A step function may be substituted as follows:

```
700 INPUT "SHIFT"; S9
702 FOR I=0 TO S9: Y(I)=0:NEXT
704 FOR I=S9 TO S9+15:Y(I)=1:NEXT
706 FOR I=S9+16 TO Q:Y(I)=0:NEXT
708 RETURN
```

5.4 STRETCHING THEOREM

The Stretching Theorem is a special, unique case of the Similarity Theorem and requires introductory comments: when dealing with real digitized data (as opposed to algebraic equations), the question arises as to how this data can be expanded. Expanding digitized data by the method we refer to as *stretching* is accomplished by simply placing zeros between the data points. The stretching theorem tells us that, if we stretch a function by placing zeros between the data points, the *spectrum* of the original function will be *repeated* in the frequency domain. Let's illustrate this diagrammatically—we represent the input data array as follows:

$$\left| \begin{array}{l} \text{DATA1} \\ \text{ARRAY} \end{array} \right| = D1,D2,D3,D4 \quad \text{----------------------} \quad (5.1)$$

Where D1,D2, etc. are the time domain data points.

Now, if we intersperse zeros between the data points:

$$\left| \begin{array}{l} \text{DATA'} \\ \text{ARRAY} \end{array} \right| = D1,0,D2,0,D3,0,D4,0 \quad \text{------------} \quad (5.2)$$

The function is now twice as long; but, except for position, it is apparent that we have not actually added any new information. At any rate, the Stretching Theorem says this; "If DATA1 ARRAY (of eqn. 5.1) has the following transform:

$$\text{Xform} \left| \begin{array}{l} \text{DATA1} \\ \text{ARRAY} \end{array} \right| = A1,A2,A3,A4 \quad \text{--------------------} \quad (5.3)$$

Where: A1,A2, etc., are the frequency domain components.

Then:

$$\text{Xform} \left| \begin{array}{l} \text{DATA'} \\ \text{ARRAY} \end{array} \right| = \tfrac{1}{2}[A1,A2,A3,A4,A1,A2,A3,A4] \quad (5.4)$$

That is, if we intersperse zeros between the data points, the frequency components A1,A2,A3,A4 simply repeat themselves a second time, with each of the component's amplitudes divided in half." Just as there was no new "data" in the time domain function, there is no new "data" in the transform. The information dealt with here is all contained in "position."

At this point it may be hard to see anything profound in this seemingly innocuous theorem; but, this simple theorem is the key to the FFT. Let's now see how it works in practice.

Run DFT5.04 and select the Stretching Theorem. The program generates a very simple waveform (four data points of +8 and -8 followed by zeros—see Fig. 5.17 below). This waveform has been selected for no other reason than that it generates a distinctive spectrum that will be easy to recognize.

T	DATA INPUT		T	DATA INPUT	
0	+8.00000	+0.00000	8	+0.00000	+0.00000
1	-8.00000	+0.00000	9	+0.00000	+0.00000
2	+8.00000	+0.00000	10	+0.00000	+0.00000
3	-8.00000	+0.00000	11	+0.00000	+0.00000
4	+0.00000	+0.00000	12	+0.00000	+0.00000
5	+0.00000	+0.00000	13	+0.00000	+0.00000
6	+0.00000	+0.00000	14	+0.00000	+0.00000
7	+0.00000	+0.00000	15	+0.00000	+0.00000

Fig. 5.17 - Un-stretched Data Input for Stretching Theorem

FREQ	F(MAG)	F(THETA)	FREQ	F(MAG)	F(THETA)
0	+0.00000	+0.00000	8	+2.00000	+0.00000
1	+0.36048	-56.25000	9	+1.81225	+33.75001
2	+0.54120	-22.50001	10	+1.30656	+67.50001
3	+0.42522	+11.25000	11	+0.63638	+101.25000
4	+0.00000	-95.71248	12	+0.00000	+95.71248
5	+0.63638	-101.25000	13	+0.42522	-11.24999
6	+1.30656	-67.50000	14	+0.54120	+22.50003
7	+1.81226	-33.74999	15	+0.36048	+56.25003

Fig 5.18 - Un-stretched Data Spectrum

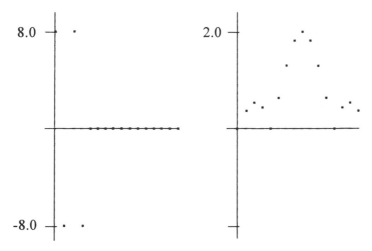

Figure 5.19 - Graphical Display of Xform F(x)

The transform of this input function is displayed in figures 5.18 and 5.19, where its distinctive "three humped" spectrum is apparent.

We now consider the "stretched" function:

T	DATA INPUT		T	DATA INPUT	
0	+8.00000	+0.00000	16	+0.00000	+0.00000
1	+0.00000	+0.00000	17	+0.00000	+0.00000
2	-8.00000	+0.00000	18	+0.00000	+0.00000
3	+0.00000	+0.00000	19	+0.00000	+0.00000
4	+8.00000	+0.00000	20	+0.00000	+0.00000
5	+0.00000	+0.00000	21	+0.00000	+0.00000
6	-8.00000	+0.00000	22	+0.00000	+0.00000
7	+0.00000	+0.00000	23	+0.00000	+0.00000
8	+0.00000	+0.00000	24	+0.00000	+0.00000
9	+0.00000	+0.00000	25	+0.00000	+0.00000
10	+0.00000	+0.00000	26	+0.00000	+0.00000
11	+0.00000	+0.00000	27	+0.00000	+0.00000
12	+0.00000	+0.00000	28	+0.00000	+0.00000
13	+0.00000	+0.00000	29	+0.00000	+0.00000
14	+0.00000	+0.00000	30	+0.00000	+0.00000
15	+0.00000	+0.00000	31	+0.00000	+0.00000

Fig. 5.20 - Stretched Input

This is the same function with the same four data points except that now they are separated by zeros (i.e. we *stretch* the function. Note that *all* of the data is "stretched"—including zeros—yielding 32 data points). The transform of this function shows the spectrum "doubling" with the amplitude components cut in half as described above. This is more readily apparent in the graphical display of fig. 5.22.

FREQ	F(MAG)	F(THETA)	FREQ	F(MAG)	F(THETA)
0	+0.00000	+0.00000	16	+0.00000	+90.00000
1	+0.18024	-56.25000	17	+0.18024	-56.25004
2	+0.27060	-22.50001	18	+0.27060	-22.49995
3	+0.21261	+11.25000	19	+0.21261	+11.25001
4	+0.00000	-95.71248	20	+0.00000	-67.38148
5	+0.31819	-101.25000	21	+0.31819	-101.25000
6	+0.65328	-67.50000	22	+0.65328	-67.50001
7	+0.90613	-33.74999	23	+0.90613	-33.75000
8	+1.00000	+0.00000	24	+1.00000	+0.00000
9	+0.90613	+33.75001	25	+0.90613	+33.75003
10	+0.65328	+67.50001	26	+0.65328	+67.50003
11	+0.31819	+101.25000	27	+0.31819	+101.25010
12	+0.00000	+0.00000	28	+0.00000	-48.59289
13	+0.21261	-11.24999	29	+0.21261	-11.24988
14	+0.27060	+22.50003	30	+0.27060	+22.50005
15	+0.18024	+56.25003	31	+0.18024	+56.25002

Fig. 5.21 - Double Spectrum of the Stretched Function

Once again, we note that this phenomenon is bilateral (i.e. if we *repeat the spectrum* a second time a "stretched" version of the function will be obtained on reconstructing! Nothing could be simpler, and the same comment holds for the FFT itself.

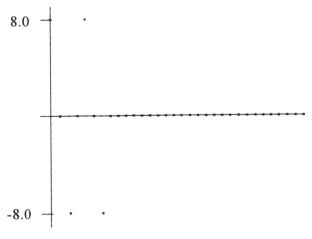

Figure 5.22 - Graphical Display of F(x)$_{\text{stretched}}$

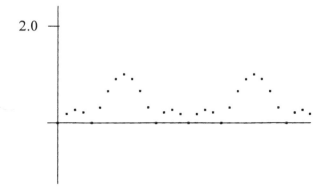

Figure 5.23 - Graphical Display of Xform F(x)$_{\text{stretched}}$

CHAPTER VI
SPEEDING UP THE DFT

6.0 INTRODUCTION

Our objective in this chapter is to reveal what makes the DFT so *unacceptably* slow. To that end we will examine individual factors that determine its "speed," and look at ways to eliminate the bottlenecks.

6.1 FUNDAMENTAL CONSIDERATIONS

First of all, just how slow is the DFT? Program DFT6.01 (listed in Appendix 6.1) is a modification of DFT5.01 allowing us to measure the time actually spent transforming data arrays of varying size. Run time vs. length of data is given below:

N	DATA POINTS	RUN TIME
4	$2^4 = 16$	5.4 SECONDS
5	$2^5 = 32$	21.8 SECONDS
6	$2^6 = 64$	88.4 SECONDS
7	$2^7 = 128$	354.7 SECONDS
8	$2^8 = 256$	1409.7 SECONDS

Fig. 6.1 - Execution Times for DFT

If we should run the program for 2^9 (= 512) data points we know the time required will be approximately 5600 seconds (1 hour and 32 minutes!) since the run time is increasing approximately as the *square* of the number of data points. You should know that, when working with the DFT, a thousand data points is not a "large" array—a two-dimensional *image,* for example, might contain a thousand lines of a thousand data points each. From the apparent trend of Fig. 6.1, however, we conclude the run time for a thousand point array would be about 6 hours—that would be 6000 hours for an entire thousand line image. That works out to be 250 days and it's only half the work required for a proper two-dimensional transform. The nature of the DFT is such that we are usually pushed toward larger data arrays, and it is easy to see that as arrays become larger processing time becomes prohibitive.

6.2 INSTRUCTION EXECUTION TIMES

Write the following routine and run it:

```
10 REM TIME TEST FOR SIN(X)
20 PI=3.14159265:E=2.7182818:P3=PI/3
22 T1=TIMER
28 FOR I = 1 TO 100
30 Y1=SIN(P3):REM LINE 30 TYP.(10 PLACES)
32 Y1=SIN(P3)
REM LINES 34 THRU 46 ARE IDENTICAL 30, 32 & 48
                   .
48 Y1=SIN(P3)
50 NEXT I
60 T1=TIMER-T1
70 PRINT T1/1000
80 STOP
```

This routine reveals the time required to evaluate a sine function. On my 80286 CPU it takes 2.95 ms (milli-seconds— 1 ms = 1×10^{-3} sec). The time required to *multiply* two numbers may be found by replacing lines 30 through 48 with:

```
30 Y1 = PI*E
```

On my computer a multiplication takes 0.482 ms. Similarly the *sum* of two numbers takes 0.421 ms. One of the reasons why these operations are so slow is that they are written in GWBASIC—there is a great deal of "overhead" that a BASIC interpreter must handle for any statement. We could make significant gains if we wrote the program in FORTRAN, or PASCAL, or C(+/++), or Assembly Language, or even compiled BASIC. With a math co-processor they will be improved even more; still, even if we slash these times by a factor of a thousand, we will not completely solve the problem—but we will come to that later.

6.3 COEFFICIENT ARRAYS

In any case we have established that finding the sine of an argument takes about $2.95/.482 = 6$ times longer than doing a multiplication. Since we require either a sine or cosine for virtually every operation we perform in the DFT, we could speed things up considerably by placing the required sine and cosine values in an array and simply calling them up when needed (as opposed to *evaluating* them each time through the loop). We can do this by modifying program DFT6.01 (see Appendix 6.2)—this modification yields the following execution times:

N	DATA POINTS		RUN TIME
4	$2^4 =$	16	2.19 SECONDS
5	$2^5 =$	32	8.51 SECONDS
6	$2^6 =$	64	33.70 SECONDS

Fig. 6.2 - Execution Times with Coefficient Array

The execution times have improved by more than a factor of two; however, if you try to run more than about 2^6 data points the computer will tell you it has run out of memory. The problem is that the array K(2,Q,Q) takes a lot of memory. We could work around this, but that's not our objective (for the moment we are only trying to find the sources of the DFT's slowness). We will ignore this limitation for the moment and continue our search for ways to speed up the DFT.

6.4 OMITTING THE NEGATIVE FREQUENCIES

As we noted back in chapter 2, when the input is limited to real numbers, we do not need to compute the negative frequencies—they are only the complex conjugates of the real frequencies. Since they eat up almost half of the time required to execute the DFT, we might well expect to double the processing speed of our DFT by eliminating them (the necessary program changes are given in Appendix 6.3). If we add these changes to our test program we obtain the following execution times:

N	DATA POINTS	RUN TIMES
4	$2^4 =$ 16	1.27 SECONDS
5	$2^5 =$ 32	4.51 SECONDS
6	$2^6 =$ 64	17.40 SECONDS

Fig. 6.3 - Execution Times (Negative Frequencies Omitted)

6.5 NON-SYMMETRICAL TRANSFORM ROUTINES

Finally, we went to a certain amount of trouble in chapter 4 to make the forward and reverse transform routines identical. This provided a very neat algorithm for our program but, in fact, it is not the fastest possible DFT routine. If we are dealing only with real number inputs, there is a certain amount of unnecessary computation going on here, and we can use the algorithms derived in the first chapters. We may add this to all of the previous modifications quite easily—we need only change the following two lines:

```
210 C(M,J)=C(M,J)+C(N,I)*K(1,J,I)
211 S(M,J)=S(M,J)+C(N,I)*K(2,J,I)
```

This will provide the forward transform only. The inverse transform will remain pretty much as it was, and will obviously have to be repeated somewhere else in the program and designated as the Inverse Transform Routine, but we are only interested in showing the improvements attainable here. If we make this change, and run the illustration one last time, we will obtain the following execution times:

N	DATA POINTS	RUN TIMES
4	2^4 = 16	0.76 SECONDS
5	2^5 = 32	2.8 SECONDS
6	2^6 = 64	10.8 SECONDS

Fig 6.4 - Execution Times (non-symmetrical Xform)

As before, the run times are shorter, but still increase as the square of the number of data points. When we consider that typical applications of the Discrete Fourier Transform handle 1024 data points and up, and even with all of our tricks, 1024 data points would still take about 45 minutes, it becomes obvious that none of these tricks will, in themselves, solve the general problem.

6.6 THE DFT COEFFICIENT MATRIX (Twiddle Factors)

I suspect that some rare disease has infiltrated science and technology within the last half century. The symptoms of this disease are the distressingly inordinate use of ACRONYMS and frivolous terminology. Thus, the term *Twiddle Factor* is an entrenched part of FFT terminology—it refers to the sine and cosine coefficients that we developed in section 6.3.

Be that as it may, the array of twiddle factors we created above (i.e. the coefficient matrix K(2,Q/2,Q)) is, in a sense, a snapshot of the DFT. If we arrange these coefficients in a matrix with the vertical positions corresponding to the *data sample times*, and the horizontal positions corresponding to the *harmonic numbers* in the frequency domain, the operation of the DFT is visible at a glance—this is a point well worth illustrating. We can write a short routine to print out this matrix:

```
10 REM DFT MATRIX
20 INPUT "NUMBER OF DATA POINTS";Q
30 DIM M(2,Q,Q/2):PI=3.14159265358#:P2=2*PI/Q
40 FOR T=0 TO Q:TP=P2*T
50 FOR F=0 TO Q/2:PRINT USING "+#.###_  ";COS(TP*F);:NEXT
60 PRINT:NEXT
99 STOP
```

This routine prints out the following matrix showing the cosine coefficients necessary to convert a 16 point database.

	F 0	F 1	F 2	F 3	F 4	F 5	F 6	F 7	F 8
T00	+1.000	+1.000	+1.000	+1.000	+1.000	+1.000	+1.000	+1.000	+1.00
T01	+1.000	+0.924	+0.707	+0.383	-0.000	-0.383	-0.707	-0.924	-1.00
T02	+1.000	+0.707	-0.000	-0.707	-1.000	-0.707	+0.000	+0.707	+1.00
T03	+1.000	+0.383	-0.707	-0.924	+0.000	+0.924	+0.707	-0.383	-1.00
T04	+1.000	-0.000	-1.000	+0.000	+1.000	-0.000	-1.000	+0.000	+1.00
T05	+1.000	-0.383	-0.707	+0.924	-0.000	-0.924	+0.707	+0.383	-1.00
T06	+1.000	-0.707	+0.000	+0.707	-1.000	+0.707	-0.000	-0.707	+1.00
T07	+1.000	-0.924	+0.707	-0.383	+0.000	+0.383	-0.707	+0.924	-1.00
T08	+1.000	-1.000	+1.000	-1.000	+1.000	-1.000	+1.000	-1.000	+1.00
T09	+1.000	-0.924	+0.707	-0.383	-0.000	+0.383	-0.707	+0.924	-1.00
T10	+1.000	-0.707	-0.000	+0.707	-1.000	+0.707	+0.000	-0.707	+1.00
T11	+1.000	-0.383	-0.707	+0.924	-0.000	-0.924	+0.707	+0.383	-1.00
T12	+1.000	+0.000	-1.000	-0.000	+1.000	+0.000	-1.000	-0.000	+1.00
T13	+1.000	+0.383	-0.707	-0.924	-0.000	+0.924	+0.707	-0.383	-1.00
T14	+1.000	+0.707	+0.000	-0.707	-1.000	-0.707	-0.000	+0.707	+1.00
T15	+1.000	+0.924	+0.707	+0.383	+0.000	-0.383	-0.707	-0.924	-1.00

Fig. 6.5 - Cosine Coefficient Matrix for 16 Point Xform

As we stated, the left hand column represents the "times" of the digitized data points, and the top row indicates the frequencies of the transformed function. Look carefully at the columns— column F0 is all ones, column F1 traces out a single cosine wave, column F2 is 2 cosine waves, etc., etc. To obtain the frequency domain function we go down each column, multiplying the time domain data points by the corresponding coefficients and summing the products.

While it has no bearing on our present development, this illustration is too good to pass up: Those who are familiar with Matrix Algebra will recognize that, if the digitized data is considered to be a Row Matrix, the process described above is nothing more than matrix multiplication. In fact, if we extend the notion to complex numbers, the whole DFT may be expressed as the product of two matrices:

$$[\ d_{11} \ d_{12} \ ... d_{1N} \ ] \quad \begin{bmatrix} c_{11} \ c_{12} \ .....c_{1N} \\ c_{21} \ c_{22} \ .....c_{2N} \\ ...... \ .. \ ...... \\ c_{N1} \ c_{N2} \ ...c_{NN} \end{bmatrix} = [\ f_{11} \ f_{12} \ ...f_{1N} \ ] \qquad .... \qquad (6.1)$$

where: C = *square matrix of coefficients*
            D = *row matrix of time domain data*
            F = *row matrix of frequency domain data*

Having noted this, we should also look at the matrix of sine coefficients. This is done by changing line 50 to print SIN(TP*F), as shown below.

	F 0	F 1	F 2	F 3	F 4	F 5	F 6	F 7	F 8
T00	+0.000	+0.000	+0.000	+0.000	+0.000	+0.000	+0.000	+0.000	+0.00
T01	+0.000	+0.383	+0.707	+0.924	+1.000	+0.924	+0.707	+0.383	-0.00
T02	+0.000	+0.707	+1.000	+0.707	-0.000	-0.707	-1.000	-0.707	+0.00
T03	+0.000	+0.924	+0.707	-0.383	-1.000	-0.383	+0.707	+0.924	-0.00
T04	+0.000	+1.000	-0.000	-1.000	+0.000	+1.000	-0.000	-1.000	+0.00
T05	+0.000	+0.924	-0.707	-0.383	+1.000	-0.383	-0.707	+0.924	-0.00
T06	+0.000	+0.707	-1.000	+0.707	-0.000	-0.707	+1.000	-0.707	+0.00
T07	+0.000	+0.383	-0.707	+0.924	-1.000	+0.924	-0.707	+0.383	-0.00
T08	+0.000	-0.000	+0.000	-0.000	+0.000	-0.000	+0.000	-0.000	+0.00
T09	+0.000	-0.383	+0.707	-0.924	+1.000	-0.924	+0.707	-0.383	-0.00
T10	+0.000	-0.707	+1.000	-0.707	-0.000	+0.707	-1.000	+0.707	+0.00
T11	+0.000	-0.924	+0.707	+0.383	-1.000	+0.383	+0.707	-0.924	+0.00
T12	+0.000	-1.000	-0.000	+1.000	+0.000	-1.000	-0.000	+1.000	+0.00
T13	+0.000	-0.924	-0.707	+0.383	+1.000	+0.383	-0.707	-0.924	-0.00
T14	+0.000	-0.707	-1.000	-0.707	-0.000	+0.707	+1.000	+0.707	+0.00
T15	+0.000	-0.383	-0.707	-0.924	-1.000	-0.924	-0.707	-0.383	-0.00

Fig. 6.6 - Sine Coefficient Matrix for 16 Point DFT

If we compare this to Fig. 6.5 we find the coefficients have changed places in an orderly manner so that now the "ones" have become "zeros" and the 0.383 values have become 0.924s, etc. If we look at the F1 column we now see that it traces out a *sine* wave, etc. The symmetry of this matrix, along with the frequent occurrence of 1, 0, and -1 brings to mind a course of action speeding up the DFT: Since multiplying by 1 or 0 requires no actual multiplication, and multiplying by -1 only requires changing the sign of the multiplicand, we might work out a scheme where much of the computation was eliminated. Furthermore, if we look across the rows we find that every value occurs at least twice. We need not repeat the operation—we could perform it once and simply place the product in the correct locations. We might do more along these lines, and perhaps make great improvements to the speed of the DFT, but most of these considerations will become irrelevant—overcome by events—once we develop the FFT. The most serious criticism of this approach, however, is that it fails to address the real problem.

Consider the above coefficient matrices overall; it is apparent that when we perform the DFT, we are performing a square matrix of mathematically identical operations. This square matrix illustrates clearly the source of our problem. It is obvious why the time of execution goes up as the square of the number of data points—the number of operations required *is* equal to the square of the number of data points.

Let's take the time to show precisely why this is such a damning characteristic: First understand that, even though your computer may run at 100 MHz, it takes perhaps a dozen clock cycles to perform a floating point *multiply and store*. Consequently, it takes dozens of clock cycles to complete a complex arithmetic multiply, sum, and store. You are *not* doing arithmetic at 100 MHz (nor, for that matter, even at 10 MHz). For the sake of argument then, let's use 1 μsec as our "bench mark" time for processing a single data point. If we have 1024 data points, we require 1,048,576 complex operations, and could process this data array in just over 1 second. This may not sound too bad, but if you are trying to design a "real time" Spectrum Analyzer, for example, you will need to do *much* better than that. Then again, it frequently turns out 1024 data points are not enough for some applications. Suppose we need 65,536 data points? This requires 4,294,967,296 operations or, at 1 μsec/operation, 4,295 seconds (71 minutes 35 seconds) to complete the transform.

Are these examples realistic? Suppose, for example, you want to process an audio signal in "real time." The audio range extends from about 20 Hz to 20,000 Hz, and you must include all of these frequencies within your data; therefore, you must digitize data for at least .05 seconds (1/20 Hz) at a rate of at least 40,000 digitizations/second. This requires a minimum array size of 2000 data points so we will use 2048 (i.e. 2^{11}). We now have 4,194,430 complex operations and a DFT time of approximately 4 seconds (at our benchmark time of 1 μsec/ operation). But, we are digitizing a new block of data every 0.05 seconds! We need to be a thousand times faster just to get the transform done. Even if we purchase a computer that's 10 times faster we will still be 100 times too slow!

The point that should be apparent from the above exercise is that, so long as the number of operations required are proportional to the square of the number of data points processed, "large" input data arrays will always (eventually) produce unmanageable processing problems.

CHAPTER VII
THE FFT

7.0 INTRODUCTION

We have come, finally, to the object of our quest—the development of the FFT algorithm proper. The objective of the FFT is simply to perform the DFT faster. As we have shown in the previous chapter, executing a DFT requires performing N^2 complex operations for N data points (NOTE: a "complex operation" includes evaluating sine and cosine functions, multiplying by the data point and adding these products to the sums of the other operations). When we realize that applications may have tens of thousands of data points (and more) we begin to understand why the conventional DFT will never suffice; 10^4 to 10^5 complex operations may be manageable, but 10^8 to 10^{10} are probably not. We need an algorithm that does for the DFT what the Horner scheme did for the series approximations of chapter 1.

How are we going to do this? Let's look at the DFT coefficient matrix again (Fig. 7.1 below). There are 8 data points and 8 harmonics (more generally, Q data points and Q frequencies), and each data point must be multiplied by its corresponding point from each harmonic's sinusoid, yielding 64 (Q^2) operations. Is there some way we can turn this liability (i.e. the number of operations being proportional to the square of the number of data points) into an asset? Well, the incurable optimist will note that if we can *reduce* the number of data points, the number of operations will be *reduced as the square*. For example, if we could split the data base into two equal parts, and process each half separately, we would have only 4 data points and 4 frequencies (i.e. 16 operations) for each half. The total would be 32 operations—only half of what we had before.

	F 0	F 1	F 2	F 3	F 4	F 5	F 6	F 7	F 8
T0	+1.000	+1.000	+1.000	+1.000	+1.000	+1.000	+1.000	+1.000	+1.000
T1	+1.000	+0.707	-0.000	-0.707	-1.000	-0.707	+0.000	+0.707	-1.000
T2	+1.000	-0.000	-1.000	+0.000	+1.000	-0.000	-1.000	+0.000	+1.000
T3	+1.000	-0.707	+0.000	+0.707	-1.000	+0.707	-0.000	-0.707	-1.000
T4	+1.000	-1.000	+1.000	-1.000	+1.000	-1.000	+1.000	-1.000	+1.000
T5	+1.000	-0.707	+0.000	+0.707	-1.000	+0.707	-0.000	-0.707	-1.000
T6	+1.000	-0.000	-1.000	+0.000	+1.000	-0.000	-1.000	+0.000	+1.000
T7	+1.000	+0.707	-0.000	-0.707	-1.000	-0.707	+0.000	+0.707	-1.000

Fig. 7.1 - Cosine Coefficient Matrix for 8 Point DFT

This, in fact, is the approach used to develop the FFT. The input data array is divided into smaller arrays to reduce the amount of computation; however, it is not clear at the outset how this can be done and still obtain the same results as provided by the DFT of the original input data. For example, if we simply split the data array in half and take the DFT of each half, we will only obtain half of the required frequency components. It is not immediately clear how the spectra of two half-sized (i.e. 4 data point) arrays might be combined to produce the 8 frequency components of the original transform.

7.1 FFT MECHANICS

The solution to our dilemma lies in the three theorems we studied in chapter 5, which we will now apply systematically to the reduction of the DFT. We start with an 8 point data array as shown below:

$$| \text{ DATA ARRAY 0 } | \; = \; | \text{ D0, D1, D2, D3, D4, D5, D6, D7 } | \qquad \text{------} \; (7.0)$$

You recall, in the addition theorem, we added two separate input functions together to form a function that was the sum of the two. Let's work that in reverse now and separate the data array shown in (7.0) into two separate arrays *that are capable of being summed to recreate the original*. There are many ways this can be done; for example, we might split the data out as follows:

$$| \text{ DATA1' } | \ = \ | \text{ D0, 0, D2, 0, D4, 0, D6, 0 } | \quad \text{---------} \quad (7.1)$$

$$| \text{ DATA2' } | \ = \ | \text{ 0, D1, 0, D3, 0, D5, 0, D7 } | \quad \text{---------} \quad (7.2)$$

It is apparent that we can add these two arrays back together to obtain the original, and from the addition theorem, we know we can add their transforms to obtain the transform of the original function. But each of these arrays has the same number of data points as the original. Each will require 64 operations to obtain its transform. If we continue along these lines we will double the amount of work rather than halving it!

Before throwing the baby out with the wash water let's take a closer look at (7.1) and (7.2). They are in the same form as the example given in the Stretching Theorem. Again, if you recall, the "stretched" data base had the unique characteristic that its transform was the same as the transform of an un-stretched data base, except it was repeated a second time (see the example for stretching in chapter 5). The light begins to dawn here, for indeed, this is the key to solving our problem. It works like this—we know that the transform of (7.1) is:

$$\text{Transform } | \text{ DATA1' } | \ = \ | \text{ F1,F2,F3,F4, F1,F2,F3,F4 } | \ \text{----} \ (7.3)$$

That is, the first four frequency components repeat a second time.

Now let's remove the zeros separating the data points in the array $|$ DATA1' $|$, and obtain the array:

$$| \text{ DATA1 } | \; = \; | \text{ D0,D2,D4,D6 } | \qquad \text{-----------------} \qquad (7.4)$$

The transform of this array is:

$$\textbf{Transform } | \text{ DATA1 } | \; = \; | \text{ F1,F2,F3,F4 } | \qquad \text{---------------} \qquad (7.5)$$

Where F1,F2,F3, and F4, in equation (7.5) are identical to F1,F2,F3 and F4 in equation (7.3). Good Grief! Can the secret of the FFT be so simple? *We obtain the transform of a stretched data array (consisting of 8 frequency components) by finding the transform of a 4 point array (4 frequencies) and repeating it!* By the Addition Theorem then, we may simply add two spectrums obtained in this manner (i.e. the transforms of $|$ DATA1' $|$ and $|$ DATA2' $|$) and we will have the transform of the original 8 point data array.   As explained above, the amount of work necessary will be halved, except that we must now include the operation of adding all of the components together.

There *is* one small problem however; the un-stretched data for $|$ DATA2' $|$ is:

$$| \text{ DATA2 } | \; = \; | \text{ D1,D3,D5,D7 } | \qquad \text{----------------------} \qquad (7.6)$$

and to get from $|$ DATA2 $|$ (eqn. 7.6) back to $|$ DATA2' $|$ (eqn. 7.2), we must not only *stretch* the data but also *shift* it one data point to the right. As we have just pointed out, to stretch the array $|$ DATA2 $|$ we simply repeat its spectrum; and, as you recall from the Shifting Theorem, if these 8 frequency components for the transform of $|$ DATA2 $|_{\text{stretched}}$ are all phase shifted (proportional to their frequency) we cause the time domain data to be *shifted*.

NOTE: Since the odd terms are shifted by only one data point we know the frequency components will be linearly phase shifted from zero (at the zero frequency component) through PI radians at the Nyquest frequency (i.e. we must shift each of the components in the DFT_{odd} transform by $2*PI*N/Q$, where N = the harmonic number of the frequency component and Q = total number of frequencies—see the Shifting Theorem in Chapter 5).

All of this is diagrammed below (Fig. 7.2)—the characteristic "crossover" pattern is sometimes called a "butterfly." The even data points are put through a 4 point DFT; the odd data points are transformed in a separate 4 point DFT. The frequency components from DFT_{odd} are properly phase shifted and summed into the frequency components of the DFT_{even}.

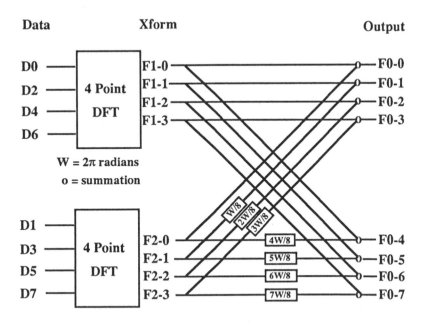

Fig. 7.2 - FFT "Butterfly" Flow Diagram

That's it! That is all there is to the basic scheme of the FFT algorithm. The presentation of the last few pages is the heart of the FFT and also the heart of this book.

Before going any further, let's see if this scheme really works. First, we need a routine that will pick out the "even" data points and perform a DFT on them. This is accomplished as follows:

As before: Q = the number of data points.
               Y(n) = the input data points.
               K1 = 2*PI/Q
               Q2 = Q/2
               J counts the frequency components.
               I counts the data points.

```
109 REM * COMPUTE EVEN DFT *
110 FOR J=0 TO Q2:J1=K1*J*2
112 FOR I=0 TO Q2-1
114 C1(J)=C1(J)+Y(2*I)*COS(J1*I)
116 S1(J)=S1(J)+Y(2*I)*SIN(J1*I)
118 NEXT I
120 C1(J)=C1(J)/Q:S1(J)=S1(J)/Q
122 NEXT J
```

Note that in line 110 the J counter only counts up to Q/2, and in line 112 the I counter counts up to (Q/2)-1. We operate on only the even data points in lines 114 and 116 by multiplying I*2 for each data point i.e. Y(I*2). Also note that we have had to multiply the argument for each sine and cosine by 2 when we define J1 at the beginning of the loop for each frequency component (line 110 ... :J1=K1*J*2). Other than these comments, we extract the standard DFT on the components; C1(J) is the cosine component for the Jth frequency term, and S1(J) is the sine component. The C1 and S1 arrays are the frequency components for the even data points. The frequency components for the odd data points will be designated C2 and S2. We perform the DFT on the odd data points in a similar routine:

```
124 REM * COMPUTE ODD DFT *
126 FOR J=0 TO Q2:J1=K1*J*2
128 FOR I=0 TO Q2-1
130 C2(J)=C2(J)+Y(2*I+1)*COS(J1*I)
132 S2(J)=S2(J)+Y(2*I+1)*SIN(J1*I)
134 NEXT I
136 C2(J)=C2(J)/Q:S2(J)=S2(J)/Q
138 NEXT J
```

This routine is identical to the one presented above except that now we only use the odd data points (lines 130 and 132— i.e. $Y(2*I+1)$).

We have now taken the transform of both halves of the data base in $2*(Q/2)^2$ (or $Q^2/2$), as opposed to Q^2 operations. We must now sum these two transforms together to obtain the complete transform. This is done as follows (in the following, K2 is defined as $K2=2*PI/Q$.):

```
139 REM * SUM HALF DFTS *
140 FC(0)=C1(0)+C2(0):FS(0)=0
142 FC(1)=C1(1)+C2(1)*COS(K2)-S2(1)*SIN(K2)
144 FS(1)=S1(1)+C2(1)*SIN(K2)+S2(1)*COS(K2)
146 FC(2)=C1(2)+C2(2)*COS(2*K2)-S2(2)*SIN(2*K2)
148 FS(2)=S1(2)+C2(2)*SIN(2*K2)+S2(2)*COS(2*K2)
150 FC(3)=C1(3)+C2(3)*COS(3*K2)-S2(3)*SIN(3*K2)
152 FS(3)=S1(3)+C2(3)*SIN(3*K2)+S2(3)*COS(3*K2)
```

The cosine term for the zero frequency component (or constant, or DC term), is found simply by adding the two cosine terms from each of the half data transforms. Since there is no sine term for zero frequency that is all there is to it. The remaining terms are slightly more complicated—the frequency components from the transform of the "odd data points" must be phase shifted before they are added in. We phase shift complex frequencies by "vector rotation," and we rotate a vector in rectangular coordinates by the operations:

$$X_{rot} = X\ Cos(A) - Y\ Sin(A) \quad\text{---------------------}\quad (7.7)$$

and;

$$Y_{rot} = X\ Sin(A) + Y\ Cos(A) \quad\text{---------------------}\quad (7.8)$$

where: A = the angle of rotation.
          X = the cosine component of the DFT.
          Y = the sine component of the DFT.

(we derive this operation in Appendix 7.1 for those who have grown "rusty.")

From this piece of information, it is apparent that (in lines 142 and on) we are rotating the frequency components obtained in the DFT for the odd data points before we sum them into the final transform.

We recognize, of course, that we have only generated Q/2 frequency components, but we require Q frequency components for the full DFT. We obtain these "latent" components by going

```
154 REM * CREATE LATENT TERMS *
156 FC(4)=C1(0)+C2(0)*COS(4*K2)-S2(0)*SIN(4*K2)
158 FS(4)=S1(0)+C2(0)*SIN(4*K2)+S2(0)*COS(4*K2)
160 FC(5)=C1(1)+C2(1)*COS(5*K2)-S2(1)*SIN(5*K2)
162 FS(5)=S1(1)+C2(1)*SIN(5*K2)+S2(1)*COS(5*K2)
164 FC(6)=C1(2)+C2(2)*COS(6*K2)-S2(2)*SIN(6*K2)
166 FS(6)=S1(2)+C2(2)*SIN(6*K2)+S2(2)*COS(6*K2)
168 FC(7)=C1(3)+C2(3)*COS(7*K2)-S2(3)*SIN(7*K2)
170 FS(7)=S1(3)+C2(3)*SIN(7*K2)+S2(3)*COS(7*K2)
```

through the two DFT arrays a second time, while continuing to shift the DFT$_{odd}$ components from PI to 2*PI radians, which is as we diagrammed it in figure 7.2 above. The reader will recognize, of course, that all of these lines of instruction (from 140 and onward) can be programmed into two loops:

```
140 FOR I=0 TO 3
142 FC(I)=C1(I)+C2(I)*COS(K2*I)-S2(I)*SIN(K2*I)
144 FS(I)=S1(I)+C2(I)*SIN(K2*I)+S2(I)*COS(K2*I)
146 NEXT I
```

and:

```
150 FOR I=4 TO 7
152 FC(I)=C1(I-4)+C2(I-4)*COS(K2*I)-S2(I-4)*SIN(K2*I)
154 FS(I)=S1(I-4)+C2(I-4)*SIN(K2*I)+S2(I-4)*COS(K2*I)
156 NEXT I
```

While this is much neater, our first objective is to understand the mechanism of the FFT—we can clean up our programming later.

The whole program is given below:

```
10 REM *** (FFT7.00A) FFT FIRST TEST ***
20 CLS:Q=8:Q2=Q/2:DIM Y(Q)
30 PI=3.141592653589793#:P2=2*PI:K1=P2/Q:K2=P2/Q
50 PRINT SPC(30);"MAIN MENU":PRINT:PRINT
60 PRINT SPC(5);"1 = ANALYZE COS COMPONENT TRIANGLE":PRINT
62 PRINT SPC(5);"2 = EXIT":PRINT
70 PRINT SPC(10);"MAKE SELECTION :";
80 A$=INKEY$:IF A$ = "" THEN 80
90 A=VAL(A$):ON A GOSUB 600,990
95 CLS: GOTO 50
100 REM *** FFT ***
105 CLS:PRINT "FREQ      F(COS)        F(SIN)":PRINT:PRINT
106 T9=TIMER
109 REM * COMPUTE EVEN DFT *
110 FOR J=0 TO Q2:J1=K1*J*2
112 FOR I=0 TO Q2-1
114 C1(J)=C1(J)+Y(2*I)*COS(J1*I)
116 S1(J)=S1(J)+Y(2*I)*SIN(J1*I)
118 NEXT I
```

```
120 C1(J)=C1(J)/Q:S1(J)=S1(J)/Q
122 NEXT J
124 REM * COMPUTE ODD DFT *
126 FOR J=0 TO Q2:J1=K1*J*2
128 FOR I=0 TO Q2-1
130 C2(J)=C2(J)+Y(2*I+1)*COS(J1*I)
132 S2(J)=S2(J)+Y(2*I+1)*SIN(J1*I)
134 NEXT I
136 C2(J)=C2(J)/Q:S2(J)=S2(J)/Q
138 NEXT J
139 REM * SUM HALF DFTS *
140 FC(0)=C1(0)+C2(0):FS(0)=0
142 FC(1)=C1(1)+C2(1)*COS(K2)-S2(1)*SIN(K2)
144 FS(1)=S1(1)+C2(1)*SIN(K2)+S2(1)*COS(K2)
146 FC(2)=C1(2)+C2(2)*COS(2*K2)-S2(2)*SIN(2*K2)
148 FS(2)=S1(2)+C2(2)*SIN(2*K2)+S2(2)*COS(2*K2)
150 FC(3)=C1(3)+C2(3)*COS(3*K2)-S2(3)*SIN(3*K2)
152 FS(3)=S1(3)+C2(3)*SIN(3*K2)+S2(3)*COS(3*K2)
154 REM * CREATE LATENT TERMS *
156 FC(4)=C1(0)+C2(0)*COS(4*K2)-S2(0)*SIN(4*K2)
158 FS(4)=S1(0)+C2(0)*SIN(4*K2)+S2(0)*COS(4*K2)
160 FC(5)=C1(1)+C2(1)*COS(5*K2)-S2(1)*SIN(5*K2)
162 FS(5)=S1(1)+C2(1)*SIN(5*K2)+S2(1)*COS(5*K2)
164 FC(6)=C1(2)+C2(2)*COS(6*K2)-S2(2)*SIN(6*K2)
166 FS(6)=S1(2)+C2(2)*SIN(6*K2)+S2(2)*COS(6*K2)
168 FC(7)=C1(3)+C2(3)*COS(7*K2)-S2(3)*SIN(7*K2)
170 FS(7)=S1(3)+C2(3)*SIN(7*K2)+S2(3)*COS(7*K2)
200 T9=TIMER-T9
210 FOR Z=0 TO Q
215 GOSUB 300
220 NEXT Z
222 PRINT:PRINT "TIME =";T9;"        ";
225 INPUT "C/R TO CONTINUE:";A$
230 RETURN
300 PRINT USING "##_    ";Z;
310 PRINT USING "+##.#####_    ";FC(Z);FS(Z)
330 RETURN
400 REM GENERATE COS COMPONENT TRIANGLE
410 FOR I=0 TO Q-1:Y(I)=0
420 FOR J=1 TO Q/2 STEP 2:Y(I)=Y(I)+COS(K1*J*I)/(J*J):NEXT
430 NEXT
440 RETURN
600 REM * COS COMPONENT TRIANGLE *
605 FOR J=0 TO Q:C1(J)=0:C2(J)=0:S1(J)=0:S2(J)=0:NEXT
610 GOSUB 400
620 GOSUB 100
630 RETURN
990 STOP
```

Figure 7.3 - Partial FFT

If we run this program we will find that the run time is noticeably less since we have reduced the total number of operations.

7.2 EXTENDING THE CONCEPT

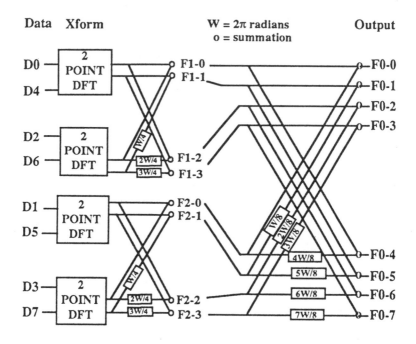

Fig. 7.4 - Double Butterfly FFT Flow Diagram

This is not the complete FFT, of course, for it is obvious that each of the *4 point DFTs* can be split into two *2 point DFTs*. We must then combine the four *2 point DFTs* into two *4 point DFTs* which are combined, as described above, into a single 8 point DFT. For our effort, we find that the total processing time will *again* be reduced by *almost* half. The flow diagram (Double Butterfly) for all this will then be as shown in Fig. 7.4 above. The mechanics of this double butterfly are essentially the same as the single butterfly described above, but we must take a moment to look at just what a two point DFT is: It will obviously have only two frequency components—a D.C. term and the Nyquest frequency. The cosine term for the zero frequency component is just the sum of the first and fourth data point divided by two (there is, of course, no sine term for a zero frequency component). The cosine term for the Nyquest frequency component is obtained by *subtracting* the fourth data point from the first, and again, there is no sine term. So, for a 2 point DFT, we have:

```
110 C3(0)=(Y(0)+Y(4))/2
112 C3(1)=(Y(0)-Y(4))/2
```

The three remaining 2 point transforms are obtained in the same manner:

```
114 C4(0)=(Y(2)+Y(6))/2
116 C4(1)=(Y(2)-Y(6))/2
118 C5(0)=(Y(1)+Y(5))/2
120 C5(1)=(Y(1)-Y(5))/2
122 C6(0)=(Y(3)+Y(7))/2
124 C6(1)=(Y(3)-Y(7))/2
```

Now we must sum these terms together to form the 4 component transforms as we did for the 8 point transform of the preceding program (FFT7.0). There is another point that we should note here—the "stretched" version of our 2 point DFT has 4 frequency components, implying 2 steps to the Nyquest frequency. Therefore, the phase shifts for the components from the "odd DFT" are multiples of PI/2 (i.e. 90°). From equations (7.7) and (7.8) we find that this special case yields:

For a shift of PI/2 $X_{rot} = -Y$ -------------- (7.9) for the first rotated component; and, for the next three components:

For a shift of PI	$X_{rot} = -X$	------------	(7.10)
For shift = 3*PI/2	$X_{rot} = Y$	------------	(7.11)
For shift = 2*PI	$X_{rot} = X$	------------	(7.12)

Similarly the rotated *sine* components are given by:

For a shift = PI/2	$Y_{rot} = X$	---------------	(7.13)
For a shift = PI	$Y_{rot} = -Y$	---------------	(7.14)
For a shift = 3*PI/2	$Y_{rot} = -X$	---------------	(7.15)
For a shift = 2*PI	$Y_{rot} = Y$	---------------	(7.16)

The summation of the 2 point frequency terms are then performed as follows:

```
126 C1(0) = (C3(0)+C4(0))/2
128 C1(1) = (C3(1))/2:S1(1)=C4(1)/2
130 C1(2) = (C3(0)-C4(0))/2:S1(2)=0
132 C1(3) = C3(1)/2:S1(3)=-C4(1)/2
134 C2(0) = (C5(0)+C6(0))/2
136 C2(1) = C5(1)/2:S2(1)=C6(1)/2
138 C2(2) = (C5(0)-C6(0))/2:S2(2)=0
140 C2(3) = C5(1)/2:S2(3)=-C6(1)/2
```

From this point we continue the FFT as we developed it for the four point DFTs in the previous section (i.e. lines beyond 140 remain the same). Let's try this as an FFT routine.

```
10 REM *** (FFT7.1) FFT 2ND TEST ***
20 Q=8:Q2=Q/2:DIM Y(Q),F(2,Q/2),K1(2,Q/4),K2(2,Q/4),Z(Q)
30 PI=3.141592653589793#:P2=2*PI:K1=P2/Q:K2=P2/Q
40 CLS:PRINT SPC(30);"MAIN MENU"
50 PRINT SPC(5);"1 = ANALYZE COS COMPONENT TRIANGLE":PRINT
60 PRINT SPC(5);"2 = EXIT ":PRINT
70 PRINT SPC(10);"MAKE SELECTION :";
80 A$=INKEY$:IF A$ = "" THEN 80
90 A=VAL(A$):ON A GOSUB 600,990
95 GOTO 40
```

```
100 REM *** FFT ***
105 CLS:PRINT "FREQ     F(COS)      F(SIN)":PRINT:PRINT
106 T9=TIMER
110 C3(0)=(Y(0)+Y(4))/2
112 C3(1)=(Y(0)-Y(4))/2
114 C4(0)=(Y(2)+Y(6))/2
116 C4(1)=(Y(2)-Y(6))/2
118 C5(0)=(Y(1)+Y(5))/2
120 C5(1)=(Y(1)-Y(5))/2
122 C6(0)=(Y(3)+Y(7))/2
124 C6(1)=(Y(3)-Y(7))/2
126 C1(0)=(C3(0)+C4(0))/2
128 C1(1)=C3(1)/2:S1(1)=C4(1)/2
130 C1(2)=(C3(0)-C4(0))/2
132 C1(3)=C3(1)/2:S1(3)=-C4(1)/2
134 C2(0)=(C5(0)+C6(0))/2
136 C2(1)=C5(1)/2:S2(1)=C6(1)/2
138 C2(2)=(C5(0)-C6(0))/2
140 C2(3)=C5(1)/2:S2(3)=-C6(1)/2
170 FOR I=0 TO Q2-1
172 FC(I)=(C1(I)+C2(I)*COS(K2*I)-S2(I)*SIN(K2*I))/2
174 FS(I)=(S1(I)+C2(I)*SIN(K2*I)+S2(I)*COS(K2*I))/2
176 NEXT
180 FOR I=Q2 TO Q-1
182 FC(I)=(C1(I-Q2)+C2(I-Q2)*COS(K2*I)-S2(I-Q2)*SIN(K2*I))/2
184 FS(I)=(S1(I-Q2)+C2(I-Q2)*SIN(K2*I)+S2(I-Q2)*COS(K2*I))/2
186 NEXT
200 T9=TIMER-T9
210 FOR Z=0 TO Q
215 GOSUB 300
220 NEXT Z
222 PRINT:PRINT "TIME =";T9
225 PRINT:PRINT:INPUT "C/R TO CONTINUE:";A$
230 RETURN
300 PRINT USING "##_    ";Z;
310 PRINT USING "+##.#####_    ";FC(Z);FS(Z)
330 RETURN
400 REM GENERATE COS COMPONENT TRIANGLE
410 FOR I=0 TO Q:Y(I)=0
420 FOR J=1 TO Q/2 STEP 2:Y(I)=Y(I)+COS(K1*J*I)/(J*J):NEXT
430 NEXT
440 RETURN
600 REM * COS COMPONENT TRIANGLE *
610 GOSUB 400
620 GOSUB 100
630 RETURN
990 STOP
```

Figure 7.5 - Double Butterfly Program Listing

It should be apparent that we performed a DFT only at the first stage of this program; the remaining stages of the process consist of simply summing the frequency components into higher order transforms at the succeeding stages. As we pointed out explicitly, a 2 point DFT is very simple; there are only cosine terms in a 2 point DFT and the cosine coefficients are either +1 or -1. There is no need to multiply at all for a 2 point DFT.

7.3 THE ONE POINT DFT

There is one last step we have to cover. If we carry the FFT scheme to its logical conclusion, we must extend the process one more step until we are dealing with 1 point DFTs. Again, we must pause to consider just what a 1 point DFT will be. There will only be one frequency component which will apparently be the zero (and/or Nyquest) frequency component. Furthermore, this zerocomponent will be the average value of the one point which is being transformed—*it is simply equal to itself!*

When we realize this, we realize that we may perform the FFT process by a shifting (i.e. rotating) and summing mechanism from the beginning—since the one point DFT is simply equal to the data point there is no necessity to perform a proper DFT at all. The full FFT algorithm, then, is illustrated in the diagram on the following page, and the first stage code will be as follows:

Noting that the shifting process will be in increments of π radians (180°), we recognize immediately that all phase shifts will require nothing more than negating terms. We negate the odd terms before adding them to the even terms to form the Nyquest frequency term for the 2 point DFT stage. The output from the first stage will be:

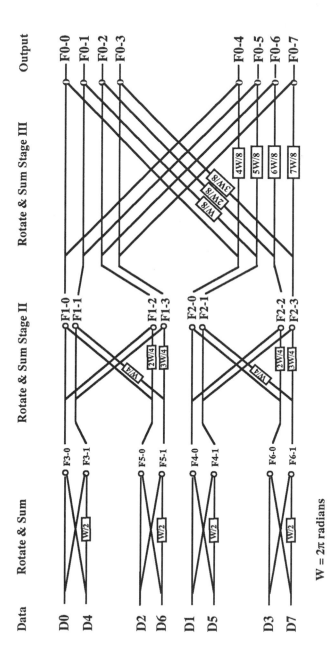

Fig. 7.6 - Full Butterfly

```
110 C3(0)=(Y(0)+Y(4))/2
112 C3(1)=(Y(0)-Y(4))/2
114 C4(0)=(Y(2)+Y(6))/2
116 C4(1)=(Y(2)-Y(6))/2
118 C5(0)=(Y(1)+Y(5))/2
120 C5(1)=(Y(1)-Y(5))/2
122 C6(0)=(Y(3)+Y(7))/2
124 C6(1)=(Y(3)-Y(7))/2
```

You will recognize this is identical to the code obtained above when we performed proper 2 point DFTs on the input data (see page 125 and/or 126). In other words there is no difference (at this low level) between rotating and summing the single point DFT equivalents and taking a DFT.

While there is no difference in a first stage *shift and add* routine vs. a first stage 2 point DFT (so far as the equations developed above are concerned), there is one *very* important difference between the two approaches. If we use a shift and add technique, we need only write a single routine which can be used to perform the shifting and summing process at every stage of the transform. We will need to work out the logistics of how the data will be handled at each stage of computation, etc., but there are a great many such considerations involved in writing a practical FFT program. These considerations will be the subject matter for our investigations in the next chapter.

One last point should be made: our objective was to reduce the number of operations required to perform a DFT. How well have we done? At *every* stage of the FFT, we now handle *all* of the data points of the data array in a shifting and adding routine. In general this routine requires half of the data points to be "rotated" which, as we have seen, requires two *complex multiplications* (p. 120). Consequently, at every stage of computation, we must perform N complex multiplications (for N data points).

Furthermore, in the scheme we have been using, $N = 2^M$, and we will have to perform M stages of shifting and adding (we will show this more clearly in the next chapter). This results in M x N operations to obtain a complete Fourier Transform (as opposed to the N^2 operations we lamented at the beginning of this chapter). Now this is not quite the "one for one" algorithm we had hoped for, but as a practical matter, it solves the problem. For example, to transform 1024 data points requires 10 x 1024 (10,240) complex operations, while a 2048 data point transform requires 11 x 2048 (22,528) complex operations (i.e. only 10% greater than a straight "one for one" increase). You can easily verify that as the number of data points increases things get better.

Based on our benchmark of 1 μsec. per operation, it would take approximately 0.022 seconds to transform the 2048 data point array we discussed in the "audio" example at the end of chapter 6—that's 0.44 seconds for transform *and* reconstruction. That doesn't leave much time to operate on the data in the transform domain, but a good engineer always has a few tricks up his sleeve (e.g. you could use *two* micro-processors and ping-pong every other block of data to the opposite processor—we now have a workable situation). In any case, this solution *is* the FFT.

CHAPTER VIII

ANATOMY OF AN FFT PROGRAM

8.0 INTRODUCTION

One might naturally assume this chapter is only a continuation of the last one; but, in fact, it deals with a completely different subject. In this chapter we assume you already know the FFT algorithm and now want to write software; this involves a completely new and unique set of problems. Here we consider the problems of data manipulation and control *within* a practical, general purpose FFT program. To the newcomer these "inner workings" of FFT software are generally considered to be nothing less than labyrinthine—and without help, they are. We hope to unscramble this subject by identifying and isolating the individual functions of these "inner workings"; still, the reader should be forewarned that this will *not* be "a piece of cake." In the past you may have spent hours trying to solve some clever puzzle of no real consequence; think of this as just such a puzzle, but one with lots of help, and of considerable significance.

We should note here that there is no single *correct* way to write an FFT program. There are countless variations, trade-offs and embellishments; in fact, there is more than one FFT algorithm. We have no intention of reviewing this mélange—*our* objective is to understand the *basic* internal operations of FFT software. To that end we will use the FFT routine of the last chapter as a "straw man", but first we will expand that program to 16 data points:

```
10 REM *** (FFT8.01) 16 POINT FFT ***
12 CLS
20 Q=16:Q2=Q/2:DIM Y(Q),FC(Q),FS(Q),KC(Q),KS(Q)
30 PI=3.141592653589793#:P2=2*PI:K1=P2/Q
40 FOR I=0 TO Q:KC(I)=COS(K1*I):KS(I)=SIN(K1*I):NEXT
50 PRINT SPC(30);"MAIN MENU":PRINT:PRINT
60 PRINT SPC(5);"1 = ANALYZE 7 COMPONENT TRIANGLE":PRINT
62 PRINT SPC(5);"2 = EXIT":PRINT
70 PRINT SPC(10);"MAKE SELECTION :";
80 A$=INKEY$:IF A$ = "" THEN 80
90 A=VAL(A$):ON A GOSUB 600,990
95 CLS:GOTO 50
99 REM  ********************
100 REM *** FFT ROUTINE ***
102 CLS:PRINT "FREQ   F(COS)       F(SIN)      ";
105 PRINT "FREQ   F(COS)      F(SIN)":PRINT:PRINT
106 T9=TIMER
108 REM *** STAGE A ***
110 A0(0)=(Y(0)+Y(8))/2
111 A0(1)=(Y(0)-Y(8))/2
112 A1(0)=(Y(1)+Y(9))/2
113 A1(1)=(Y(1)-Y(9))/2
114 A2(0)=(Y(2)+Y(10))/2
115 A2(1)=(Y(2)-Y(10))/2
116 A3(0)=(Y(3)+Y(11))/2
117 A3(1)=(Y(3)-Y(11))/2
118 A4(0)=(Y(4)+Y(12))/2
119 A4(1)=(Y(4)-Y(12))/2
120 A5(0)=(Y(5)+Y(13))/2
121 A5(1)=(Y(5)-Y(13))/2
122 A6(0)=(Y(6)+Y(14))/2
123 A6(1)=(Y(6)-Y(14))/2
124 A7(0)=(Y(7)+Y(15))/2
125 A7(1)=(Y(7)-Y(15))/2
126 REM *** STAGE B ***
127 BC0(0)=(A0(0)+A4(0))/2
128 BC0(1)=A0(1)/2:BS0(1)=A4(1)/2
129 BC0(2)=(A0(0)-A4(0))/2
130 BC0(3)=A0(1)/2:BS0(3)=-A4(1)/2
131 BC1(0)=(A1(0)+A5(0))/2
132 BC1(1)=A1(1)/2:BS1(1)=A5(1)/2
133 BC1(2)=(A1(0)-A5(0))/2
134 BC1(3)=A1(1)/2:BS1(3)=-A5(1)/2
135 BC2(0)=(A2(0)+A6(0))/2
136 BC2(1)=A2(1)/2:BS2(1)=A6(1)/2
137 BC2(2)=(A2(0)-A6(0))/2
138 BC2(3)=A2(1)/2:BS2(3)=-A6(1)/2
139 BC3(0)=(A3(0)+A7(0))/2
140 BC3(1)=A3(1)/2:BS3(1)=A7(1)/2
141 BC3(2)=(A3(0)-A7(0))/2
142 BC3(3)=A3(1)/2:BS3(3)=-A7(1)/2
```

```
148 REM *** STAGE C ***
150 FOR I=0 TO 3:J=2*I
151 CC1(I)=(BC0(I)+BC2(I)*KC(J)-BS2(I)*KS(J))/2
152 CC2(I)=(BC1(I)+BC3(I) *KC(J)-BS3(I)*KS(J))/2
153 CS1(I)=(BS0(I)+BC2(I)*KS(J)+BS2(I)*KC(J))/2
154 CS2(I)=(BS1(I)+BC3(I)*KS(J)+BS3(I)*KC(J))/2
155 NEXT I
160 FOR I=4 TO 7:J=2*I:K=I-4
161 CC1(I)=(BC0(K)+BC2(K)*KC(J)-BS2(K)*KS(J))/2
162 CC2(I)=(BC1(K)+BC3(K)*KC(J)-BS3(K)*KS(J))/2
163 CS1(I)=(BS0(K)+BC2(K)*KS(J)+BS2(K)*KC(J))/2
164 CS2(I)=(BS1(K)+BC3(K)*KS(J)+BS3(K)*KC(J))/2
165 NEXT I
168 REM *** STAGE F ***
170 FOR I=0 TO Q2-1
172 FC(I)=(CC1(I)+CC2(I)*KC(I)-CS2(I)*KS(I))/2
174 FS(I)=(CS1(I)+CC2(I)*KS(I)+CS2(I)*KC(I))/2
176 NEXT
180 FOR I=Q2 TO Q-1
182 FC(I)=(CC1(I-Q2)+CC2(I-Q2)*KC(I)-CS2(I-Q2)*KS(I))/2
184 FS(I)=(CS1(I-Q2)+CC2(I-Q2)*KS(I)+CS2(I-Q2)*KC(I))/2
186 NEXT
200 T9=TIMER-T9
210 FOR Z=0 TO Q2-1
215 GOSUB 300
220 NEXT Z
222 PRINT:PRINT "TIME =";T9
225 PRINT:PRINT:INPUT "C/R TO CONTINUE:";A$
230 RETURN :REM *** END FFT ROUTINE ***
235 REM      ***************************
300 PRINT USING "##";Z;:PRINT "   ";
310 PRINT USING "+##.####";FC(Z);:PRINT "   ";
312 PRINT USING "+##.####";FS(Z);:PRINT "     ";
320 PRINT USING "##";Z+Q2;:PRINT "   ";
322 PRINT USING "+##.####";FC(Z+Q2);:PRINT "    ";
324 PRINT USING "+##.####";FS(Z+Q2)
330 RETURN
400 REM GENERATE 7 COMPONENT TRIANGLE
410 FOR I=0 TO Q:Y(I)=0
420 FOR J=1 TO Q/2 STEP 2:Y(I)=Y(I)+COS(K1*J*I)/(J*J):NEXT
430 NEXT
440 RETURN
600 REM * 7 COMPONENT TRIANGLE *
610 GOSUB 400: REM GENERATE INPUT FUNCTION
620 GOSUB 100: REM XFORM FUNCTION
630 RETURN
990 STOP
```

Figure 8.0 - 16 Point FFT

8.1 STAGES OF COMPUTATION (BUTTERFLIES)

Look closely at lines 108 to 186 of the computer listing on the previous two pages (Fig. 8.0). This FFT routine is constructed as a series of *stages of computation* (A, B, C and F). While it is apparent that each of these stages is different, we know from the previous chapter they are, in a sense, doing the same thing (performing a butterfly). This observation is the foundation of this chapter; we must sort out what is the same, and what is different, in each of these *stages*. We can, of course, write a single iterative loop to perform the parts that are the same. This loop must then be nested within another loop (or loops) which can change the parts that are different between stages (and applications e.g. data array size). This simple methodology, applied to the algorithm of the last chapter, is all there is to understanding this "difficult" subject. Let's begin by reviewing this basic structure—let's recall why we wrote this program in *stages of computation* in the first place.

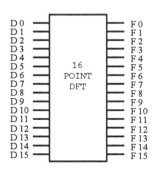

Fig. 8.1 - 16 Point DFT

NOTE: The following discussion concerns only the *data flow* within the FFT. We will discuss the "butterflies" themselves shortly, but for now we simplify things by representing the computation that takes place within a DFT as a simple box (Fig. 8.1).

1. At the outset, our objective is to find the Discrete Fourier Transform for an array of data designated D(x). For this particular example x takes on all values from 0 to 15, but in general we will consider any array size that is an integer power of 2 (i.e. 2^n).

2. Before we begin the 16 point DFT though, we realize we can speed things up by splitting this 16 point array into two 8 point arrays. We split out every other data point so that one array will have all of the odd data points and the other will have all of the even. By taking the DFTs of two 8 point arrays we will reduce the work by almost half.

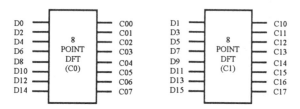

Fig. 8.2 - 8 Point DFTs

3. We recognize, of course, that we can speed things up even more by splitting each of these 8 point arrays into 4 point arrays.

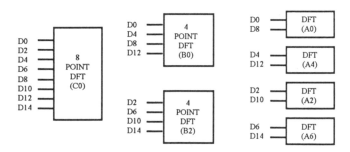

Fig. 8.3 - Breakdown of "Even" Data Points

4. The general scheme is apparent by now—we can split each 4 point DFT into 2 point DFTs, and these can be split into "1 point DFTs." In Fig. 8.3 above we show this breakdown for the *even* data points (the same thing must happen for the odd data points of course). None of this shows up in the program code however, it is only a mental exercise. It is from this point forward that we write our program—a shifting and adding "butterfly" that starts from *one point DFTs* and works back up through

stages of "butterflies" (see Fig. 8.4). Fig. 8.3 only shows how the data is broken out, and consequently, how it must be recombined in succeeding *stages of computation.*

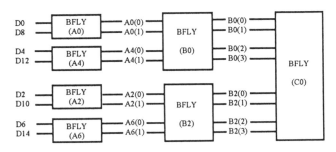

Fig. 8.4 - "Even Data" Stageing for FFT

5. We perform the same process for the odd data points, of course (Fig. 8.5). These two 8 point DFTs (i.e. C0 above and C1 below), will then be combined in a final butterfly using an identical shifting and adding process to form the equivalent of the original 16 point DFT.

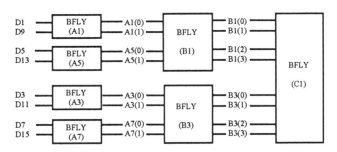

Fig. 8.5 - "Odd Data" Stageing for FFT

These diagrams contain nothing but butterflies. We know, of course, that the butterflies are identical in function—only the number of components handled (and magnitude of phase shifts) is different. Surely we can write a routine to accommodate all of these stages (or *any* number of stages) if we can only find a set of rules describing exactly how each succeeding stage differs from its predecessor. Let's point out some general rules that are already apparent:

1. The first stage of computation will have an input array of Q (Q = 2^n) data points each of which will be treated as a single point DFT. The second stage will have inputs from Q/2 partial DFTs and each partial DFT will have 2 frequency components. The third stage will have Q/4 partial (4 component) DFTs, etc., etc. Each stage will halve the number of partial DFTs, and double the number of frequency components (total number of "elements" remaining a constant = Q).

2. Therefore, since we start with Q partial DFTs, and halve the number at each stage (until we have a single DFT), an array of 2^n data points must yield n *stages of computation* in the FFT!

3. Note that in the first stage, the components summed in the butterflies (both odd and even) are all separated by exactly 1/2 the length of the data array! Whether this is so for the following stages of computation (or not) depends on how we handle the data.

There is more to be gleaned from the general flow of data shown in Figs. 8.2 through 8.5, but its significance might not yet be so obvious; so, we will move forward at this time and consider the butterflies.

8.2 MECHANICS OF THE BUTTERFLY

The butterflies are the simplest imaginable routines—we add one complex number to another (phase shifted) complex number. The only *question* here is: "How much is each component shifted at each stage of computation?" From chapter 5 we know that a one data point shift (in the time domain) will cause a linearly increasing phase shift (in the frequency domain) resulting in 180° shift at the Nyquest frequency (continuing on linearly to the end of the array). Also, from the discussion of the previous chapter (section 7.1) we know that each time we split a data array, the resulting "odd" element array will be shifted by *one data point* (the "even" element array is *not* shifted of course). For the "A" stage output there are only two frequency components—the zero frequency component and the

Nyquest frequency component. Therefore, these two components will receive, respectively, phase shifts of zero and 180° (i.e. we multiply the higher order data point by -1 when forming the Nyquest frequency—see Fig. 8.0, stage A, lines 108-125), resulting in the following:

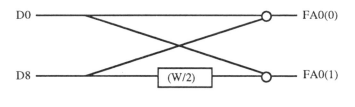

W = 2 π radians shift

Fig. 8.6 - Stage "A" Shifting and Adding (Typical)

The notation (W/2) indicates that 1/2 of a full cycle (180°) of phase shift is performed; FA0(0) indicates the zero frequency component out of the A0 butterfly—the "○" indicates a "summing junction" where the lines converging from the left are *summed* (complex quantities).

The phase shift at stage B will be:

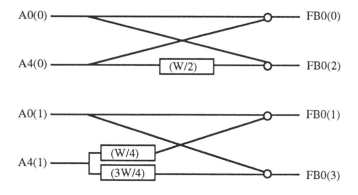

Fig. 8.7 - Stage "B" Shifting (Typical)

Note that, in Fig. 8.7, inputs A4(0) and A0(1) have been reversed in position to make the diagram simpler. In the computer program A0(0) is added to A4(0), then A0(1) is added to A4(1) shifted by (W/4). Next the virtual components will be created by adding A0(0) to A4(0)*(W/2) and A0(1) to A4(1)*(3W/4) [here "*" = "shifted by"]. Also note that while the frequency components were rotated by *increments* of 1/2 of a cycle (i.e. PI radians) in stage "A," they are rotated by increments of 1/4 cycle (PI/2 radians) in stage "B." Apparently they will be rotated by increments of 1/8 cycle (i.e. PI/4 radians) in stage "C," etc., etc. Let's add this to our set of rules:

4. The "phase shift increment" (i.e. the incremental amount that each successive component's phase is *increased*) starts at PI (i.e. 180°) for the first stage and halves for each stage thereafter until the nth stage (where it will be $PI/2^{(N-1)} = 2\pi/Q$).

We may also note the following at this time:

5. It is an almost trivial observation, but we will always add the 0 frequency component of one butterfly to the 0 frequency component of another, 1st component to 1st, etc., etc. This is true for both the direct components as well as the "latent components."

6. Finally, when we sum two arrays together (such as A0 and A4, or B1 and B3 back in Fig. 8.4 and 8.5), we will "rotate" or phase shift only the components from the "time shifted arrays" (i.e. A4 and B3).

From these six observations, we may now write not only a general purpose butterfly routine, but most of the FFT program. Let's state explicitly how we intend to do this:

Since, from the foregoing, we know the "arithmetic" of the butterflies will be identical for each stage of computation (i.e. shifting and adding complex numbers), we will make this operation our core *iterative routine*. On the other hand the phase shift, number of partial DFTs, and number of components in a partial DFT will change for every stage. To control these variables, we will use a "stage control loop" around our "universal butterfly." For each *stage of computation*, this control loop will change the number of DFTs, number of components within each DFT, and phase shifts—all according to the rules just established. This shouldn't be too difficult then, but before we actually write our FFT routine, it will be prudent to consider how, exactly, we are going to manage the data within the data arrays. Let me explain:

8.3 ARRAY DATA MANAGEMENT

In the program listed at the beginning of this chapter (Fig. 8.0), each stage of computation has a *separate data array* to save the data generated at that stage. Now, all of these arrays are no problem when dealing with 16 data points, but, as the number of data points increases, and array size increases, this inefficient use of memory is disastrous.

It is possible to write a program with only one complex quantity array; the two data points (complex) to be added are pulled from the array, rotated, and combined (twice) creating two new pieces of data. The new data is then stored back into the locations from whence the original data was taken (intermediate results are saved in temporary storage until a place is available in the array). This is the most efficient use of memory possible; however, it is *inefficient in execution* since it requires multiple data transfers for every operation, slowing things down.

A practical compromise may be accomplished with two arrays which are "ping-ponged" between the output and input from stage to stage. This is the approach we will use, and it will be explained in detail below, but the reader will recognize this is a design "trade-off" decision.

This is certainly simple enough; but, if we follow standard practice, and manage data as outlined two paragraphs back, the *final* data will be out of sequence in the array. An additional "stage" will therefore be required (usually referred to as "bit reversal") to sort the data points into proper order. This is a hangover from using a single data array, and as we have said, *we will not do things that way in our FFT routine*. We will bite the bullet and face *data management* from the outset, and avoid the wasted time used up by this bit reversal "stage."

So, to save memory and facilitate the fastest possible execution, we will use two separate "2 degree of freedom" arrays [DIMensioned as $C(2,Q)$ and $S(2,Q)$]. To control these arrays we will use a pair of complementary switches (T0 designates which side of the array inputs data to the butterfly and T1 designates the side receiving output). After each stage of computation we reverse the "1" and "0" states of T0 and T1, effectively causing the array sides to "ping-pong" (i.e. they reverse their input/output functions). This brings us to a crucial consideration: at each stage of computation, the data stored in the output array must be in proper position when it is used as input (by the same routine) for the next stage of computation. Now, we pointed out that the *first stage* sums the data from the 0th position in the array with the data in the $Q/2$th position (Rule 3); the 1st position with the $Q/2+1$ position; etc.; etc. We will be well advised, then, to add the following to our set of rules:

> 7. DEFINITION 1 - We may store data in the array at any convenient location, so long as the data is *managed* in such a way that any two components which must be summed together (both in the present and the next stage of computation) will be separated by half the length of the array.

What advantage does this provide? Well, returning to stage 1 computation, data points $D(0)$ and $D(Q/2)$ are used twice—first to form the $A0(0)$ component, and then $A0(1)$. We are no longer forced to store these components back into the 0 and $Q/2$ positions in the array; in fact, we will put them into the *first two locations* of the output array! The two components to which *they* must be added in the next stage of computation (i.e. $A4(0)$ and $A4(1)$ in figure 8.4), however, *must* be placed in locations $Q/2$ and $Q/2+1$. Continuing on, D1 and D9 will combine to form $A1(0)$ and $A1(1)$, which we place in position 2 and 3 of the output array. If we continue in this manner, Definition 1 is satisfied (see Fig. 8.8 below).

0	D00		A0(0)	DFT
1	D01		A0(1)	A0
2	D02	STAGE	A1(0)	DFT
3	D03	"A"	A1(1)	A1
4	D04	COMPUTATION	A2(0)	DFT
5	D05	("A" BUTTERFLIES)	A2(1)	A2
6	D06		A3(0)	DFT
7	D07		A3(1)	A3
8	D08	T0 = 0 T1 = 1	A4(0)	DFT
9	D09		A4(1)	A4
10	D10		A5(0)	DFT
11	D11		A5(1)	A5
12	D12		A6(0)	DFT
13	D13		A6(1)	A6
14	D14		A7(0)	DFT
15	D15		A7(1)	A7

INPUT ARRAY   OUTPUT ARRAY

Fig. 8.8 - Input/Output Data Position in the
First Stage of Computation for a 16 Point FFT

Note that the data arrangement of both input and output arrays
are in accordance with the requirements of figures 8.4, 8.5, and
also Definition 1 above. We must follow the same format in stage
"B" of the computation, of course, as shown in figure 8.9 below.

0	A0(0)		B0(0)	
1	A0(1)		B0(1)	DFT
2	A1(0)	STAGE	B0(2)	B0
3	A1(1)	"B"	B0(3)	
4	A2(0)	COMPUTATION	B1(0)	
5	A2(1)	("B" BUTTERFLIES)	B1(1)	DFT
6	A3(0)		B1(2)	B1
7	A3(1)		B1(3)	
8	A4(0)	T0 = 1 T1 = 0	B2(0)	
9	A4(1)		B2(1)	DFT
10	A5(0)		B2(2)	B2
11	A5(1)		B2(3)	
12	A6(0)		B3(0)	
13	A6(1)		B3(1)	DFT
14	A7(0)		B3(2)	B3
15	A7(1)		B3(3)	

INPUT ARRAY   OUTPUT ARRAY

Fig. 8.9 - Data Position Between Input and
Output Arrays for Second Stage of Computation

We keep each partial DFT *together*. The *DFTs* in the output array at each stage are arranged so that when used for the *input array in the next stage* of computation, we still sum the 0th element with the 8th element; the 1st with the 9th; etc.

0	B0(0)		C0(0)	
1	B0(1)		C0(1)	
2	B0(2)	STAGE	C0(2)	
3	B0(3)	"C"	C0(3)	DFT
4	B1(0)	COMPUTATION	C0(4)	
5	B1(1)	("C" BUTTERFLIES)	C0(5)	C0
6	B1(2)		C0(6)	
7	B1(3)		C0(7)	____
8	B2(0)	T0 = 0 T1 = 1	C1(0)	
9	B2(1)		C1(1)	
10	B2(2)		C1(2)	
11	B2(3)		C1(3)	DFT
12	B3(0)		C1(4)	
13	B3(1)		C1(5)	C1
14	B3(2)		C1(6)	
15	B3(3)		C1(7)	

INPUT ARRAY   OUTPUT ARRAY

Fig. 8.10 - Data Position for Input and
Output Arrays for 3rd Stage of Computation

0	C0(0)		F0(0)	
1	C0(1)		F0(1)	
2	C0(2)	STAGE	F0(2)	
3	C0(3)	"F"	F0(3)	
4	C0(4)	COMPUTATION	F0(4)	
5	C0(5)	("F" BUTTERFLY)	F0(5)	FINAL
6	C0(6)		F0(6)	
7	C0(7)		F0(7)	DFT
8	C1(0)	T0 = 1 T1 = 0	F0(8)	
9	C1(1)		F0(9)	
10	C1(2)		F0(10)	
11	C1(3)		F0(11)	
12	C1(4)		F0(12)	
13	C1(5)		F0(13)	
14	C1(6)		F0(14)	
15	C1(7)		F0(15)	

INPUT ARRAY   OUTPUT ARRAY

Fig. 8.11 - Data Position in Final Input and Output Arrays

8.4 THE FFT ALGORITHM

Rather than presenting a completed program with all of the above considerations neatly incorporated, let's develop our algorithm by introducing changes sequentially (insofar as possible) into the program given at the beginning of this chapter. Our first order of business will be to throw out all of the different data arrays and introduce the data array structure discussed above.

Reviewing lines 108 to 186 in FFT8.01 (Fig. 8.0) we note that stages A and B are written *in line*, while stages C and F employ *loops*. In our *Universal FFT routine* we hope to employ a single, iterative loop for all of the stages; so, it will be best to start with stage F. Recall that in stage F we are summing the two 8 point partial DFTs into the final 16 component DFT (see Fig. 8.12 below).

```
168 REM *** STAGE F ***
170 FOR I=0 TO Q2-1
172 FC(I)=(CC0(I)+CC1(I)*KC(I)-CS1(I)*KS(I))/2
174 FS(I)=(CS0(I)+CC1(I)*KS(I)+CS1(I)*KC(I))/2
176 NEXT
180 FOR I=Q2 TO Q-1
182 FC(I)=(CC0(I-Q2)+CC1(I-Q2)*KC(I)-CS1(I-Q2)*KS(I))/2
184 FS(I)=(CS0(I-Q2)+CC1(I-Q2)*KS(I)+CS1(I-Q2)*KC(I))/2
186 NEXT
```

Fig. 8.12 - Stage "F" Butterfly

You should be familiar with this routine by now; we form the F stage components by summing the C0 stage components with the *rotated* C1 components. We essentially perform the same loop twice: once rotating the C1 components from 0 to π radians (i.e. the "direct" components in lines 170-176), then once again rotating from π to 2π (lines 180-186, creating the "latent" components). Okay, let's start the modifications by changing out the data arrays:

```
180 REM *** STAGE "F" BUTTERFLY ***
184 FOR I=0 TO Q3:REM Q2 = Q/2 AND Q3 = Q2-1
187 C(0,I)=(C(1,I)+C(1,I+Q2)*KC(I)-S(1,I+Q2)*KS(I))/2
188 S(0,I)=(S(1,I)+C(1,I+Q2)*KS(I)+S(1,I+Q2)*KC(I))/2
190 NEXT I
192 FOR I=Q2 TO Q-1
194 C(0,I)=(C(1,I-Q2)+C(1,I)*KC(I)-S(1,I)*KS(I))/2
195 S(0,I)=(S(1,I-Q2)+C(1,I)*KS(I)+S(1,I)*KC(I))/2
197 NEXT I
```

Fig. 8.13 - 1st Modification of Stage "F" Butterfly

As we said, the data is now handled in arrays dimensioned as $C(2,Q)$ and $S(2,Q)$. Recognizing that, when we enter stage "A" the data will be in the $C(\underline{0},x)$ side of the cosine component array, the *output* data for stage A will be placed in the $C(\underline{1},x)$ and $S(\underline{1},x)$ arrays. In the "B" stage computations, the input data is *taken* from the $\underline{1}$ side and output to the $\underline{0}$ side. These two "sides" of the data array *ping-pong* back and forth as we proceed through the succeeding stages (see Fig. 8.8 through 8.11) until, in stage F as shown above, the data is input from the $\underline{1}$ side and output to the $\underline{0}$ side. Also note that the data is *managed* as shown in Fig. 8.11 in keeping with rule 3 (p. 137), and Definition 1 (p. 141).

Let's look quickly at the operation of this loop (in the following $Q =$ total number of data points, $Q2 = Q/2$ and $Q3 = Q2-1$). The *C0 DFT* data is located in the lower half addresses of the input arrays and the *C1 DFT* data is in the upper half (see Fig. 8.11). We access the C1 data by adding Q2 to I in lines 187 and 188; the C0 data is accessed in lines 194 and 195 by *subtracting* Q2 from I. Otherwise the routine is the same as shown in Fig. 8.12.

So far so good—this is how we will handle data array management, but the "F" stage considered alone is just a little *too* easy. We know, in general, that each succeeding stage will halve the number of partial DFTs and double the number of frequency components (rule 1). We need some way to track this within the data array, but this gets us into the general arena of *things that change between stages*. Let's look at how we handle this whole area of concern by modifying the "C" stage:

```
180 REM ***  STAGE "C" BUTTERFLY ***
182 QT=Q/4:KT1=2
184 FOR J=0 TO Q3 STEP 4:J1=2*J:K9=J+Q2
185 FOR I=0 TO QT-1:KT=I*KT1:K=K9+I
187 C(1,J1+I)=(C(0,I+J)+C(0,K)*KC(KT)-S(0,K)*KS(KT))/2
188 S(1,J1+I)=(S(0,I+J)+C(0,K)*KS(KT)+S(0,K)*KC(KT))/2
190 NEXT I
191 J1 = J1+QT
192 FOR I=0 TO QT-1:KT=(I+QT)*KT1:K=K9+I
194 C(1,J1+I)=(C(0,I+J)+C(0,K)*KC(KT)-S(0,K)*KS(KT))/2
195 S(1,J1+I)=(S(0,I+J)+C(0,K)*KS(KT)+S(0,K)*KC(KT))/2
197 NEXT I:NEXT J
```

Figure 8.15 - Modified Stage "C"

We immediately define QT (which stands for QuiT) as the number of frequency components in any partial DFT (for this stage QT=4). KT1 relates to the *stage dependent* phase shift increment—KT the address of the sine and cosine twiddle factors (for this stage, KT1 = 2). We will talk about this shortly.

Next, note that we have nested the "I" loop, which constituted stage F, inside a "J" *control loop* (good lord, they're actually going to do what they said). However, while the equations of this new "I" loop are of the same form as in stage "F", addressing the elements of the data arrays is considerably more complex. Let's look closely at the operation, keeping in mind that in stage "C" we form the *C0 DFT* by summing *DFT B0* and *DFT B2* (half an array away), and the *C1* by summing *B1* and *B3*: As before, lines 185 through 190 generate the "direct" components, and lines 192 through 197 generate the "latent" components. At line 185 "I" will step from 0 to QT-1 (as we noted above, the first 4 components, which constitute *B0*). KT locates the correct twiddle factor, but now we use K to generate the address of the *B2 DFT* components in the upper half of the array. K satisfies our requirement to sum components separated by half an array (i.e. Q/2) via lines 184 (where we set K9 equal to J+Q/2), and line 185 (where we set K

= K9+I). At the start J=0 so that K will equal I+Q/2. This results in (lines 187 and 188) our summing the components located at I (J = 0) with the rotated components located at I+Q/2. Note that we place these sums into the output side of our two sided array at a location of I + J1, and if you look back at line 184 you will find that J1 = 2*J = 0. This "I" loop will sum the first 4 components in the array with the first 4 components half an array away and place the results in the first 4 locations of the output array. It then moves on to generate the 4 latent components.

The "I" loop for the latent components is almost (but not quite) identical. We know that we will store these results just above the direct components in the output array, so we solve *this* requirement by adding QT to J1 in line 191. Also, to generate the latent components, we must *continue increasing* the rotations of the high order partial DFT components, and we do this by adding QT to I (in line 192) before multiplying by KT1. Otherwise...identical.

What about the "J" loop? You will have already figured out that J keeps track of the partial DFTs (*B0/B2* and *B1/B3* in this stage). When we have finished creating the *C0 DFT* in the output array (i.e. the sum of *B0* and *B2* as just described) we increment J by 4 (line 184) and execute the two I loops again. We now see the purpose of J1 and K9; J1 doubles J and is the *partial DFT index* for the output array, while K9 adds J to Q/2 to provide an input array *partial DFT index*.

That's it...

It is apparent that this routine for the "C" stage is more general than that presented earlier for the "F" stage. In fact, you will note that if we set QT = Q/2, KT1 = 1, and the STEP increment for the J loop = 8, this "C" loop will work for the "F" loop also. *In fact, with very little effort, it can be made to work for all of the stages!* Before we do that, however, let's clear up a little detail we have left dangling since back in section 8.2.

8.5 TWIDDLE FACTOR INDEXING

In lines 30 and 40 of FFT8.01 (Fig. 8.0, p. 132) we generate the twiddle factors. These are the sine and cosine values we will use to rotate the complex numbers of one DFT before we add them to the complex numbers of another [see. (7.7) and (7.8), p. 120)]. Now, by rule 4 (p. 139), the nth stage will require every value generated in this twiddle factor table. The N-1 stage of computation, however, will only require every second value from this table (but for *two* partial DFTs). The N-2 stage requires every 4th twiddle factor, etc., until we reach the first stage of computation which only requires the first value and the middle of the table value. This is what the twiddle factor index, KT1 specifies. It is multiplied by I or (I+QT) for the actual twiddle factor address KT.

8.6 THE COMPLETE FFT

From this point, the next step in the development of a general purpose FFT routine is more or less obvious: The whole FFT routine may

```
181 REM *** FFT ROUTINE ***
182 FOR M=0 TO N-1:QT=2^M:KT1=2^(N-M-1)
183 REM *** "UNIVERSAL" BUTTERFLY ***
184 FOR J=0 TO Q3 STEP QT:J1=2*J:K9=J+Q2
185 FOR I=0 TO QT-1:KT=I*KT1:K=K9+I
187 C(T0,J1+I)=(C(T1,I+J)+C(T1,K)*KC(KT)-S(T1,K)*KS(KT))/2
188 S(T0,J1+I)=(S(T1,I+J)+C(T1,K)*KS(KT)+S(T1,K)*KC(KT))/2
190 NEXT I
191 J1 = J1+QT
192 FOR I=0 TO QT-1:KT=(I+QT)*KT1:K=K9+I
194 C(T0,J1+I)=(C(T1,I+J)+C(T1,K)*KC(KT)-S(T1,K)*KS(KT))/2
195 S(T0,J1+I)=(S(T1,I+J)+C(T1,K)*KS(KT)+S(T1,K)*KC(KT))/2
196 NEXT I:NEXT J
197 IF T0=0 THEN T0=1:T1=0:GOTO 199
198 T0=0:T1=1
199 NEXT M
```

Figure 8.16 - Complete FFT routine

be compressed into a single stage, which will be repeated as many times as necessary to perform the complete transform (Fig 8.16). We do this by nesting the previous routine (Fig. 8.15) inside an "M loop," where M counts the *stages of computation*. This loop sets the value of QT (partial DFT size) and KT1 (skip index for twiddle factors) at the beginning of each pass. M counts up from 0 to N-1 (i.e. the number of stages that must be executed), and QT may be calculated simply by figuring 2^M (i.e. 2^m), which obviously starts at 1 and doubles for each pass through the loop. KT1 is calculated by finding the value of $2^{((N-1)-M)}$, which starts at $2^{(N-1)}$ [i.e. $2^{(4-1)} = 8$ for the 16 point FFT of this chapter] and then halves at each stage until it becomes 1 (i.e. the final pass through the loop).

We have inserted our toggle (T0 and T1) to select the input and output sides of the data array. At the end of the "M loop" (lines 197 and 198) we reverse these values to switch the input and output registers.

To use this routine in the program presented at the beginning of this chapter (see below, Fig. 8.17) we must make a few changes. In line

```
10 REM *** (FFT8.02) Q=2^N POINT FFT ***
12 PRINT "INPUT NUMBER OF DATA POINTS AS 2^N"
14 INPUT "N = ";N
16 Q=2^N
20 Q2=Q/2:Q3=Q2-1:Q4=Q/4:Q5=Q4-1:Q8=Q/8
22 DIM Y(Q),C(2,Q),S(2,Q),KC(Q),KS(Q)
30 PI=3.141592653589793#:P2=2*PI:K1=P2/Q
32 FOR I=0 TO Q:KC(I)=COS(K1*I):KS(I)=SIN(K1*I):NEXT
40 CLS
50 PRINT SPC(30);"MAIN MENU":PRINT:PRINT
60 PRINT SPC(5);"1 = ANALYZE Q/2 COMPONENT TRIANGLE":PRINT
64 PRINT SPC(5);"2 = EXIT":PRINT
70 PRINT SPC(10);"MAKE SELECTION :";
80 A$=INKEY$:IF A$ = "" THEN 80
90 A=VAL(A$):ON A GOSUB 600,990
95 GOTO 40
100 REM *** FFT ***
102 CLS:PRINT "FREQ   F(COS)      F(SIN)      ";
105 PRINT "FREQ   F(COS)      F(SIN)":PRINT:PRINT
106 T9=TIMER
181 REM *** FFT ROUTINE ***
182 FOR M=0 TO N-1:QT=2^M:KT1=2^(N-M-1)
```

```
183 REM *** "UNIVERSAL" BUTTERFLY ***
184 FOR J=0 TO Q3 STEP QT:J1=2*J:K9=J+Q2
185 FOR I=0 TO QT-1:KT=I*KT1:K=K9+I
187 C(T0,J1+I)=(C(T1,I+J)+C(T1,K)*KC(KT)-S(T1,K)*KS(KT))/2
188 S(T0,J1+I)=(S(T1,I+J)+C(T1,K)*KS(KT)+S(T1,K)*KC(KT))/2
190 NEXT I
191 J1 = J1+QT
192 FOR I=0 TO QT-1:KT=(I+QT)*KT1:K=K9+I
194 C(T0,J1+I)=(C(T1,I+J)+C(T1,K)*KC(KT)-S(T1,K)*KS(KT))/2
195 S(T0,J1+I)=(S(T1,I+J)+C(T1,K)*KS(KT)+S(T1,K)*KC(KT))/2
196 NEXT I:NEXT J
197 IF T0=0 THEN T0=1:T1=0:GOTO 199
198 T0=0:T1=1
199 NEXT M
200 T9=TIMER-T9
210 FOR Z=0 TO Q2-1
215 GOSUB 300
220 NEXT Z
222 PRINT:PRINT "TIME =";T9
225 PRINT:PRINT:INPUT "C/R TO CONTINUE:";A$
230 RETURN
300 PRINT USING "###";Z;:PRINT "    ";
310 PRINT USING "+##.#####";C(T1,Z);:PRINT "    ";
312 PRINT USING "+##.#####";S(T1,Z);:PRINT "       ";
320 PRINT USING "###";Z+Q2;:PRINT "    ";
322 PRINT USING "+##.#####";C(T1,Z+Q2);:PRINT "    ";
324 PRINT USING "+##.#####";S(T1,Z+Q2)
330 RETURN
400 REM GENERATE Q/2 COMPONENT TRIANGLE
410 FOR I=0 TO Q:C(0,I)=0:S(0,I)=0
420 FOR J=1 TO Q/2 STEP 2:C(0,I)=C(0,I)+COS(K1*J*I)/(J*J):NEXT
430 NEXT
440 RETURN
600 REM * Q/2 COMPONENT TRIANGLE *
602 CLS:PRINT:PRINT
604 PRINT "PREPARING DATA INPUT - PLEASE WAIT!"
610 GOSUB 400
612 T0=1:T1=0
614 PRINT:INPUT "DATA READY - C/R TO CONTINUE";A$
620 GOSUB 100
630 RETURN
990 STOP
```

Fig. 8.17 - Listing for FFT8.02

420 we now input the test function data to the 0 side of the C(0,Q) array, and set the S(0,Q) side to zero. To print the results out, we must determine which side of the array was used to receive the last pass through the loop. This is determined automatically, since the output is always placed in the side that T0 is set to; however, we toggle T0 one last time as we exit the final pass through the loop, and consequently we will print the side indicated by T1.

We can now arbitrarily select the number of data points that we wish to handle in our FFT, and we allow the selection of N at the beginning of the program before we dimension the arrays. We do this in lines 12 through 16 and the job is finished. The complete program is shown above in Fig 8.17. Type it into your computer and try it. If the program works, and you understand it all, it's okay to feel a little pride (it's justifiable). In fact, you deserve a break—perhaps even a beer and pizza. Unfortunately, that's out of the question. We still have unfinished business—another "mile to go before we sleep" so to speak.

8.5 THE INVERSE FFT

We already have the inverse FFT of course—it's the same algorithm we just developed (see chapter 4). We need only make a change in "scale factors", and some minor changes in program control. Let's talk about the scale factor first.

When we take the forward transform we must multiply through by a unit amplitude sinusoid, sum up all the products, and then divide by the number of terms to find the *average value* of the resultant. When we reconstruct the time domain function however, we only need to sum up all of the points from the individual sinusoids—no averaging is involved. Consequently, if we use the identical routine for both the forward *and* reverse transforms, there will be a scale factor error of Q in one or the other. To correct for this discrepancy we must introduce a *scale factor term* into the routine. We can set this term to either Q or 1.0, effectively scaling the results for a forward or inverse transform.

In the actual butterflies of the program we have just presented (Fig 8.17, lines 187, 188, 194 and 195), the right hand sides of these equations are all divided by 2. This division by 2 for M stages of computation achieves the scale factor requirement (i.e. Q) for the forward transform, but is the culprit that introduces an error into the results of an inverse transform. If we replace the "2" in these equations with a variable "SK1", we may set SK1 to 2 when doing the forward transform and to 1 when doing the inverse. The changes required will be:

```
187 C(T0,J1+I)=(C(T1,I+J)+C(T1,K)*KC(KT)-S(T1,K)*KS(KT))/SK1
188 S(T0,J1+I)=(S(T1,I+J)+C(T1,K)*KS(KT)+S(T1,K)*KC(KT))/SK1

194 C(T0,J1+I)=(C(T1,I+J)+C(T1,K)*KC(KT)-S(T1,K)*KS(KT))/SK1
195 S(T0,J1+I)=(S(T1,I+J)+C(T1,K)*KS(KT)+S(T1,K)*KC(KT))/SK1
```

and at line 612 we set SK1 = 2 when we set the T0 and T1 flags:

```
612 T0=1:T1=0:SK1=2
```

This takes care of the forward transform. To achieve the inverse transform we must make the following changes to the "MAIN MENU":

```
10 REM *** (FFT8.10A) FFT/INV FFT ***
  .
  .
  .
62 PRINT SPC(5);"2 = INVERSE TRANSFORM":PRINT
64 PRINT SPC(5);"3 = EXIT":PRINT
90 A=VAL(A$):ON A GOSUB 600,700,990
```

Then, at line 700, we write the inverse transform routine.

```
700 REM *** INVERSE TRANSFORM ***
710 SK1 = 1
712 CLS:PRINT "TIME    AMPLITUDE    NOT USED        ";
714 PRINT "TIME    AMPLITUDE    NOT USED":PRINT:PRINT
720 GOSUB 106
730 RETURN
```

We set SK1 to 1 (line 710) thereby removing the forward transform scale factor, and print a new heading for the output data. Having done this we GOSUB to line 106 of the FFT routine. When we return the inverse FFT will have been completed.

Put these changes into the program developed previously and run the forward and inverse transform for N=4. Then compare the results to those obtained in the previous chapters.

Okay, you've earned that break now.

AUTHOR'S CLOSING REMARKS

As is always true, when you completely understand a presentation, you will completely understand its limitations. There are a multitude of shortcomings and places for easy improvement in the software presented in this chapter... and they will remain there as challenges—as exercises. If improvements are not immediately apparent you might want to review chapter 6, or even re-read this one. That is the advantage (and curse) of writing your own software—you can always find ways to improve it.

Copies of this book can be purchased at your local bookstore or, if unavailable there, ordered directly from the publisher. Please include $2.75 shipping and handling—Florida residents add 6% sales tax.

Citrus Press

P.O. Box 10062

Indian River Station

Titusville, FL 32783

APPENDIX 1.1

BASIC Programming Language

BASIC is simple to use and easy to learn. If you want to multiply two numbers together, you type:

```
Y = 2*2 (The * replaces x and must always be used
to indicate multiplication.)
```

If you want to divide two numbers you type:

```
Y = 4/2  (The / sign indicates division just as in
algebra.)
```

To add or subtract you type:

```
Y = 4+2  or Y = 4-2
```

Nothing could be simpler! [NOTE: Formally, all of the above statements are written: LET Y = 2*2; LET Y = 4/2; etc. The LET term is simply gold plating and usually omitted by experienced programmers. We will drop it from the start.]

BASIC, of course, is more useful than that. You don't have to type the actual numbers in, you can type:

```
Y=A*B: PI = 3.14: Y1=2*PI+A*B: or AREA = R*R*3.14
```

You can also type:

```
Y=SIN(3.14); or Y=COS(N*PI); or Z=ATN(Y/X)
[ATN=ArcTaNgent]
```

and, of course, much more. Note that sometimes we use spaces between the characters and sometimes not. When writing equations the spaces are optional and BASIC ignores them. This is not true when writing KEY WORDS such as LET, COS, FOR, THEN, etc., etc. Spaces are required following KEY WORDS.

To "program" in BASIC you simply write down a sequence of instructions; the computer will perform them one step after the other. All you have to do is number the lines so the computer will know in what order to perform the steps:

```
10 PI = 3.14159265358: E = 2.7182818
20 Y = SIN(E*PI)
30 PRINT Y
40 END
```

Note that in line 10 we typed two instructions—all we had to do was separate them with a colon. In line 30 we introduced a new instruction PRINT. PRINT Y will cause the value of Y to be displayed on the screen. Displaying the result of a computation is *not* automatic—we do not waste computer time displaying intermediate results. Line 40 tells the computer to stop running the program—this isn't really necessary in this case since BASIC would have stopped anyway when it ran out of instructions. Typing STOP or END *is* necessary sometimes, and always a good habit. NOTE: Some versions of BASIC do not require line numbers, but GWBASIC requires them and we are trying to keep things simple.

What can we really do with BASIC? Suppose we want to simulate the input to a hypothetical digital system. Suppose, for example, the input is normally created by running a sine wave generator into an A/D converter (A/D = Analog to Digital). The A/D "samples" the sine wave at regular intervals and presents 16 digitized "words" to the computer for every cycle of sine wave. How can we simulate this?

```
10 N=0: PI=3.14159: K1=PI/8
20 Y=SIN(N)
30 PRINT Y
40 N=N+K1
50 GOTO 20
```

The new instruction, GOTO 20, obviously causes the computer to go back to line 20 where it continues executing the instructions. There is no END statement here though—this program *never* ends. It is an endless "loop", and that could be a problem...

There are two ways to get out of this "endless loop":

1) change the ending of the routine to this:

```
50 IF N < 2*PI THEN 20
60 STOP: END
```

The IF ... THEN statements gives us a great deal of control. It allows the computer to make decisions. Line 50 now reads: If N is less than 2 times PI (i.e. N< 2*PI) THEN go back to line 20. If this condition is not met the execution simply goes straight ahead to the next instruction which will stop and end the routine. We could also have written:

```
50 IF N=> PI-1.14159*1024 THEN 70
60 GOTO 20
70 STOP: END
```

This routine will run quite a bit longer than the previous one. In line 60 the instruction GOTO is oneword. [Note: You may recognize a potential problem here. Does the computer multiply before subtracting or does it perform the sequence as it is written? If you are uncertain, group things with parentheses ((PI-1.14159)*1024).]

BASIC is generally pretty flexible in how we use the instructions. The second way to get around the endless loop problem is:

2) change the routine to look like this:

```
10 PI=3.14159: K1=PI/8
12 FOR N = 0 TO 2*PI STEP K1
20 Y=SIN(N)
30 PRINT Y
40 NEXT N
50 STOP: END
```

The FOR statement (line 12) must always be used with a NEXT statement (line 40)—they come as a pair. This statement is designed specifically for making loops, and does quite a bit for us: a) It tells the computer that the following instructions (until it encounters a NEXT statement) are part of a loop. b) It defines a "loop counter" (variable N in the above example). c) It gives the starting value for the loop counter (in this case 0). d) It gives the ending value of the loop counter (2*PI). e) It also gives the increment by which the loop counter is to increase each time we step through the loop (STEP K1). The STEP term is optional and is 1.0 if not stated otherwise. All in all, this is quite a bit for a single instruction.

Line 40 has been changed to a NEXT N statement which tells the computer to go back to the beginning of the loop (i.e. line 12). The N following the NEXT is, like LET, unnecessary; but, in this case, it helps us keep track when we start "nesting" one loop inside of another.

Notice that we sort of slipped the FOR statement in between lines 10 and 20. GWBASIC is nice about that—you can type in a new line at any time and BASIC will put it in the correct place for you (*you* must specify the line number, of course). If you want to delete a line, simply type the line number and hit the Enter (Carriage Return) key.

I/O (Input/Output)

The PRINT statement used above is an output command. It "outputs" something to an "output device" (i.e. the display screen). The disk drive and printer are other examples of output devices. We also need to input things to a program occasionally, and this is frequently done from the keyboard. In BASIC we use the INPUT statement. For example:

```
INPUT "NUMBER OF TERMS";N
```

causes the computer to print **NUMBER OF TERMS?** to the display screen (as a prompt for the user) and then stop and wait for a number to be typed into the keyboard. What it is *really waiting* for is a Carriage Return to be typed, so when we have typed the desired number we hit the carriage return to let the computer know we have finished "inputting" the necessary data. The computer then takes the number that was typed and assigns it to the variable N. Neat!

That's enough for now. You can read the program in chapter 1. We will take the time to explain all new instructions as we go along.

[NOTE: There are, of course, a great many instructions, techniques, and rules that we have not mentioned here. We will cover what we need as we go along; but, there will be a great deal that we *never* mention. Be forewarned that at the beginning you will probably find it slow going when you read the programs in this book. Don't be discouraged—very shortly you will be "sight reading" them.]

APPENDIX 5.1

The programs of chapter 5 are based on a single "core" with specific Execution Control Subroutines used to illustrate the individual theorems. This is essentially the same program developed in preceding chapters and, consequently, we will discuss only the modifications here.

```
4 REM   ********************************************
6 REM   ** (DFT5.00A) GENERATE/ANALYZE WAVEFORM **
8 REM   ********************************************
10 Q=32
12 PI=3.141592653589793#:P2=2*PI:K1=P2/Q:K2=1/PI
14 DIM C(2,Q),S(2,Q),KC(2,Q),KS(2,Q)
16 CLS:FOR J=0 TO Q:FOR I=1 TO 2:C(I,J)=0:S(I,J)=0:NEXT:NEXT
20 CLS:REM *    MAIN MENU     *
22 PRINT:PRINT:PRINT "        MAIN MENU":PRINT
24 PRINT " 1 = THEOREM ILLUSTRATION":PRINT
31 PRINT " 2 = EXIT":PRINT:PRINT
32 PRINT SPC(10);"MAKE SELECTION";
34 A$ = INKEY$:IF A$="" THEN 34
36 A=VAL(A$):ON A GOSUB 300,1000
38 GOTO 20
40 CLS:N=1:M=2:K5=Q:K6=-1:GOSUB 108
42 FOR J=0 TO Q:C(2,J)=0:S(2,J)=0:NEXT
44 GOSUB 200: REM - PERFORM DFT
46 GOSUB 140: REM - PRINT OUT FINAL VALUES
48 PRINT:INPUT "C/R TO CONTINUE";A$
50 RETURN
80 CLS:GOSUB 150:REM PRINT HEADING
81 FOR I=0 TO Q-1:C(1,I)=0:S(1,I)=0:NEXT
82 N=2:M=1:K5=1:K6=1
84 GOSUB 200:REM INVERSE TRANSFORM
86 GOSUB 140:REM PRINT OUTPUT
88 PRINT:INPUT "C/R TO CONTINUE";A$
90 RETURN
100 REM  ****************************************
101 REM  *        PROGRAM SUBROUTINES          *
102 REM  ****************************************
104 REM  *      PRINT COLUMN HEADINGS          *
105 REM  ****************************************
106 REM  *    FREQUENCY DOMAIN HEADING         *
107 REM  ****************************************
108 PRINT:PRINT :IF COR$="P" THEN 116
109 PRINT "FREQ   F(COS)  F(SIN)   FREQ    F(COS)   F(SIN)"
110 PRINT
111 RETURN
112 REM  ****************************************
113 REM  *      POLAR COORDINATES HEADING      *
114 REM  ****************************************
```

```
116 PRINT "FREQ   F(MAG)  F(THETA) FREQ   F(MAG)  F(THETA)"
118 GOTO 112
137 REM ******************************
138 REM *       PRINT OUTPUT      *
139 REM ******************************
140 IF COR$="P" AND M=2 THEN GOSUB 170
141 FOR Z=0 TO Q/2-1
142 PRINT USING "##_    ";Z;
144 PRINT USING "+###.#####_    ";C(M,Z),S(M,Z);
145 PRINT USING "##_    ";(Z+Q/2);
146 PRINT USING "+###.#####_    ";C(M,Z+Q/2),S(M,Z+Q/2)
147 NEXT Z
148 RETURN
150 REM *****************************************
152 REM *   PRINT TIME DOMAIN COLUMN HEADINGS   *
153 REM *****************************************
154 PRINT
156 PRINT "                    RECONSTRUCTION":PRINT
158 PRINT " T                          T":PRINT
160 RETURN
169 REM  ***************************************************
170 REM  * CONVERT FROM RECTANGULAR TO POLAR COORDINATES  *
171 REM  ***************************************************
172 FOR I=0 TO Q-1
174 MAG=SQR(C(M,I)^2+S(M,I)^2)
175 IF C(M,I)=0 THEN 190
176 ANGLE =180/PI*ATN(S(M,I)/C(M,I))
177 IF C(M,I)>0 THEN S(M,I)=ANGLE:GOTO 180
178 IF ANGLE>0 THEN S(M,I)=ANGLE-180
179 IF ANGLE<0 THEN S(M,I)=ANGLE+180
180 C(M,I)=MAG:NEXT
182 RETURN
190 IF S(M,I)=0 THEN 180
192 S(M,I)=90:GOTO 180
200 REM ******************************
202 REM *    TRANSFORM/RECONSTRUCT   *
204 REM ******************************
206 FOR J=0 TO Q-1:REM SOLVE EQNS FOR EACH FREQUENCY
208 FOR I=0 TO Q-1:REM MULTIPLY AND SUM EACH POINT
210 C(M,J)=C(M,J)+C(M,I)*COS(J*I*K1)+K6*S(N,I)*SIN(J*I*K1)
211 S(M,J)=S(M,J)-K6*C(N,I)*SIN(J*I*K1)+S(N,I)*COS(J*I*K1)
212 NEXT I
214 C(M,J)=C(M,J)/K5:S(M,J)=S(M,J)/K5:REM SCALE RESULTS
216 NEXT J
218 RETURN
220 REM ******************************
222 REM *       PLOT FUNCTIONS      *
224 REM ******************************
225 SFF=16:SFT=64
226 SCREEN 9,1,1,1:COLOR 9,1,1:CLS:YF=-1:YT=-1
228 LINE (0,5) - (0,155):LINE (0,160)- (0,310)
230 LINE (0,155) - (600,155):LINE (0,310)-(600,310)
```

```
232 GOSUB 266 :REM SET SCALE FACTORS
234 COLOR 15,1,1
236 FOR N=0 TO Q-1 :REM PLOT DATA
238 GOSUB 260 :REM CONVERT DATA TO PIXELS
240 LINE (X,Y) - (X,Y):LINE (X,Z)-(X,Z)
242 NEXT N
244 LOCATE 2,10:PRINT "FREQUENCY DOMAIN (MAG)"
246 LOCATE 14,12:PRINT "TIME DOMAIN"
248 LOCATE 24,1
250 INPUT "C/R TO CONTINUE";A$
252 SCREEN 0,0,0
254 RETURN
256 REM ******************************
257 REM * COMPUTE SCREEN LOCATIONS  *
258 REM ******************************
260 Y=SQR(C(2,N)^2+S(2,N)^2):Y=155-(YF*Y)
261 X=N*600/Q:Z=310-(YT*C(1,N))
262 RETURN
263 REM ******************************
264 REM *   SET & PRINT SCALE FACTORS *
265 REM ******************************
266 YF=150/SFF:YT=150/SFT:LINE (0,5)-(5,5):LINE (0,80)-(5,80)
268 LINE (0,160)-(5,160):LINE (0,235)-(5,235)
270 LOCATE 1,2:PRINT SFF :LOCATE 6,2:PRINT SFF/2
272 LOCATE 12,2:PRINT SFT:LOCATE 17,2:PRINT SFT/2
274 RETURN
```

Figure A5.1 - Core Program Listing

In line 10 we now define a variable Q. In a general purpose program we will need the capability to select the length of the input function. In this program we provide that capability by defining Q to be the length of the input data array. In line 12 we define K1 in terms of Q and in line 14 we dimension all of the data arrays in terms of Q. In line 16 we initialize the primary data arrays by setting up the loop in terms of Q. This sort of reorienting of the program loops in terms of Q takes place throughout the program of course.

At line 170 we provide a routine to convert from rectangular to polar coordinates. This is done by finding the RSS (square root of the sum of the squares) of the cosine and sine components. The angle is found as the arctangent of the sine divided by the cosine component (the answer is converted from radians to degrees). A certain amount of overhead is required (lines 175 - 179) to determine in which quadrant the angle lies (BASIC assumes angles are in the first and fourth quadrants—

we must make the determination for a true four quadrant system).

Starting at line 220 we provide a routine to plot the two functions. It may be interesting to review this routine if you have never plotted anything on the screen before; otherwise, it is hardly worth the effort, since different versions of BASIC have different methods of plotting data. In GWBASIC we must select the "screen" we want to work in (there are several), and the colors to be used for foreground and background (see line 226). Lines 228 and 230 draw the X-Y coordinates. We then jump down to line 266 to set the scale factors and print this data on the screen (this will be different for each illustration). We then return to line 234 where we change the color of our plotting "foreground" and begin to plot the functions (lines 236 through 242). When plotting points on the screen (in GWBASIC) the points are located in a matrix of "pixels" and the data of the functions plotted must be converted to these coordinates. The actual locations depends on the "screen" selected, the computer being used, and version of BASIC. If you are using GWBASIC with any PC/CLONE with VGA, this routine will work fine—otherwise there may be problems. The data to be plotted is transformed into screen coordinates in the subroutine at lines 260 - 262.

APPENDIX 5.2

The following routines are presented essentially without comment. They are the four routines that modify the program of Figure A5.1 above to perform the demonstrations of the Theorems. They are relatively simple and should be readable with the help of the remarks.

```
4 REM    *****************************************
6 REM    ** (DFT5.01A) GENERATE/ANALYZE WAVEFORM **
24 PRINT " 1 = SIMILARITY THEOREM":PRINT
299 REM ******************************
300 CLS:REM *   SIMILARITY THEOREM    *
302 REM CLEAR ARRAYS
304 FOR I=0 TO Q-1:C(1,I)=0:S(1,I)=0
306 FOR J=1 TO 2:KC(J,I)=0:KS(J,I)=0:NEXT:NEXT
308 CLS:PRINT "WIDTH =";F9:REM DISPLAY CURRENT WIDTH
```

```
310 INPUT "WIDTH ";F9 :REM INPUT WIDTH
311 REM CHECK WIDTH LIMITS
312 IF F9>Q/2 THEN PRINT Q/2;" DATA POINTS MAXIMUM":F9=Q/2
314 IF F9<1 THEN PRINT "1 DATA POINTS MINIMUM":F9=1
316 PRINT SPC(13);"SIMILARITY TEST - WIDTH =";F9
317 FOR I=Q/2-F9 TO Q/2+F9:REM GENERATE INPUT FUNCTION
318 C(1,I)=Q*(SIN(PI*(I-(Q/2-F9))/(2*F9)))^2
319 NEXT
320 GOSUB 158:REM PRINT HEADING
322 M=1:GOSUB 140:REM PRINT INPUT FUNCTION
324 PRINT:INPUT "C/R TO CONTINUE";A$
326 GOSUB 40:REM TAKE XFORM
328 GOSUB 220:REM PLOT DATA
330 PRINT "MORE (Y/N)?";
332 A$=INKEY$:IF A$="" THEN 332
334 IF A$="Y" OR A$="y" THEN 304
396 RETURN
1000 STOP
```

Figure A5.2 - DFT5.01 - Similarity Theorem

```
4 REM   ******************************************
6 REM   ** (DFT5.02A) GENERATE/ANALYZE WAVEFORM **
24 PRINT " 1 = ADDITION THEOREM":PRINT
299 REM *********************************
300 CLS:REM *    ADDITION THEOREM    *
301 REM CLEAR DATA ARRAYS
302 FOR I=0 TO Q-1:C(1,I)=0:S(1,I)=0
304 FOR J=1 TO 2:KC(J,I)=0:KS(J,I)=0:NEXT:NEXT
308 REM *** GENERATE ADDITION TEST FUNCTION ***
310 PRINT:PRINT SPC(13);"EXPONENTIAL RISING EDGE":PRINT
312 FOR I=0 TO Q/2-1:C(1,I)=1-EXP(-I/5):NEXT
314 GOSUB 158: REM PRINT HEADING
316 M=1:GOSUB 140: REM PRINT INPUT FUNCTION
318 PRINT:INPUT "C/R TO CONTINUE";A$
320 GOSUB 40: REM TAKE XFORM
322 GOSUB 220: REM PLOT DATA
323 REM SAVE RISING EDGE TRANSFORM
324 FOR I=0 TO Q-1:KC(1,I)=C(1,I):KC(2,I)=C(2,I):KS(2,I)=S(2,I):NEXT
326 FOR I=0 TO Q-1:C(1,I)=0:NEXT
328 PRINT:PRINT SPC(13);"EXPONENTIAL FALLING EDGE":PRINT
330 K4=1-EXP(-Q/10) :REM SET INITIAL VALUE
332 FOR I=Q/2 TO Q-1:C(1,I)=K4*EXP(-(I-(Q/2))/5):NEXT
334 GOSUB 158:REM PRINT HEADING
336 M=1: GOSUB 140:REM PRINT INPUT DATA
338 PRINT:INPUT "C/R TO CONTINUE";A$
340 GOSUB 40:REM TRANSFORM DATA
```

```
341 GOSUB 220:REM PLOT DATA
342 CLS
343 PRINT "SUM XFORMS OF RISING AND FALLING EXPONENTIAL FUNCTIONS"
344 FOR I=0 TO Q-1:C(2,I)=C(2,I)+KC(2,I):S(2,I)=S(2,I)+KS(2,I):NEXT
345 M=2:PRINT:GOSUB 108:REM PRINT HEADING
346 GOSUB 140:REM PRINT SUM OF XFORMS
347 PRINT:INPUT "C/R TO CONTINUE";A$
348 CLS:GOSUB 150:REM PRINT HEADING
349 GOSUB 81:REM INVERSE TRANSFORM W/O HEADING
350 GOSUB 220:REM PLOT DATA
351 REM * SUM BOTH RISING AND FALLING TIME DOMAIN FUNCTIONS *
353 PRINT:PRINT SPC(10);"EXPONENTIAL RISING EDGE";
354 FOR I=0 TO Q/2-1:C(1,I)=1-EXP(-I/5):NEXT
356 PRINT " + EXPONENTIAL FALLING EDGE":PRINT
358 K4=1-EXP(-Q/10) :REM SET INITIAL VALUE
360 FOR I=Q/2 TO Q-1:C(1,I)=K4*EXP(-(I-(Q/2))/5):NEXT
362 GOSUB 158:REM PRINT HEADING
364 M=1: GOSUB 140:REM PRINT INPUT DATA
366 PRINT:INPUT "C/R TO CONTINUE";A$
368 GOSUB 40:REM TRANSFORM DATA
370 GOSUB 220:REM PLOT DATA
372 RETURN
```

Figure A5.3 - DFT5.02 - Addition Theorem

```
6 REM   ** (DFT5.03A) GENERATE/ANALYZE WAVEFORM **
24 PRINT " 1 = SHIFTING THEOREM":PRINT
299 REM ******************************
300 CLS:REM *    SHIFTING THEOREM    *
301 REM ******************************
302 FOR I=0 TO Q-1:C(1,I)=0:S(1,I)=0
304 FOR J=1 TO 2:KC(J,I)=0:KS(J,I)=0:NEXT:NEXT
305 COR$="P"
308 REM *** GENERATE IMPULSE FUNCTION ***
310 PRINT:PRINT SPC(18);"IMPULSE FUNCTION":PRINT
312 C(1,0)=32
314 GOSUB 158:REM PRINT HEADING
316 M=1:GOSUB 140:REM PRINT INPUT FUNCTION
318 PRINT:INPUT "C/R TO CONTINUE";A$
320 GOSUB 40:REM TAKE XFORM
322 GOSUB 220:REM PLOT DATA
324 FOR I=0 TO Q-1:C(1,I)=0:S(1,I)=0:NEXT
326 INPUT "AMOUNT OF SHIFT (0-31)";S9
328 C(1,S9)=32
330 GOSUB 158:REM PRINT HEADING
332 M=1:GOSUB 140:REM PRINT INPUT FUNCTION
334 PRINT:INPUT "C/R TO CONTINUE";A$
```

```
336 GOSUB 40:REM TAKE XFORM
338 GOSUB 220:REM PLOT DATA
340 PRINT "CONTINUE ILLUSTRATION ? (Y/N)"
342 A$=INKEY$:IF A$="" THEN 342
344 IF A$="Y" OR A$="y" THEN 324
346 RETURN
```

Figure A5.4 - DFT5.03 - Shifting Theorem

```
4 REM   *******************************************
6 REM   ** (DFT5.04A) GENERATE/ANALYZE WAVEFORM **
24 PRINT " 1 = STRETCHING THEOREM": PRINT
299 REM ********************************
300 CLS : REM *   STRETCHING THEOREM   *
301 REM ********************************
302 FOR I = 0 TO Q - 1: C(1, I) = 0: S(1, I) = 0
304 FOR J = 1 TO 2: KC(J, I) = 0: KS(J, I) = 0: NEXT: NEXT
305 COR$ = "P": Q = 16: K1 = P2 / Q
306 GOSUB 900
308 REM *** GENERATE "Z1" FUNCTION ***
310 PRINT : PRINT SPC(18); " - Z1 - FUNCTION": PRINT
312 C(1, 0) = 8: C(1, 1) = -8: C(1, 2) = 8: C(1, 3) = -8
314 GOSUB 158: REM PRINT HEADING
316 M = 1: GOSUB 140: REM PRINT INPUT FUNCTION
318 PRINT : INPUT "C/R TO CONTINUE"; A$
320 GOSUB 40: REM TAKE XFORM
322 GOSUB 220: REM PLOT DATA
324 FOR I = 0 TO Q - 1: C(1, I) = 0: S(1, I) = 0: NEXT
326 Q = 32: K1 = P2 / Q
328 C(1, 0) = 8: C(1, 2) = -8: C(1, 4) = 8: C(1, 6) = -8
330 GOSUB 158: REM PRINT HEADING
332 M = 1: GOSUB 140: REM PRINT INPUT FUNCTION
334 PRINT : INPUT "C/R TO CONTINUE"; A$
336 GOSUB 40: REM TAKE XFORM
338 GOSUB 220: REM PLOT DATA
396 RETURN
```

Figure A5.5 - DFT5.04 - Stretching Theorem

APPENDIX 5.3
PROOF OF THE THEOREMS

In the following proofs the fundamental definition of the Fourier Transform is taken as:

$$F(f) = \int_{-\infty}^{\infty} f(t)\, e^{-i2\pi ft}\, dt \quad \text{-----------------------} \quad (A5.3.1)$$

THE SIMILARITY THEOREM

The Similarity theorem states: If $F(f)$ is the Fourier transform of $f(t)$, then the transform of $f(at)$ will be $F(f/a)/|a|$ ($|a|$ = magnitude of a). This follows directly from (A5.3.1) when $f(at)$ is substituted for $f(t)$:

$$\int_{-\infty}^{\infty} f(at)\, e^{-i2\pi ft}\, dt = \frac{1}{|a|} \int_{-\infty}^{\infty} f(at)\, e^{-i2\pi(f/a)(at)}\, d(at)$$

$$= \frac{1}{|a|}\, F(f/a) \quad \text{------------------} \quad (A5.3.2)$$

Note: Replacement of t with (at) in the exponential term requires that we replace f with (f/a).

THE ADDITION THEOREM (LINEARITY)

The theorem is: If $F(f)$ and $G(f)$ are the transforms of $f(t)$ and $g(t)$ respectively then $F(f)+G(f)$ will be the transform of $f(t)+g(t)$. Again, this results from a direct application of the integral of (A5.3.1):

$$\int_{-\infty}^{\infty} [f(t)+g(t)]e^{-i2\pi ft}dt = \int_{-\infty}^{\infty} f(t)e^{-i2\pi ft}dt \quad + \quad \int_{-\infty}^{\infty} g(t)e^{-i2\pi ft}dt$$

$$= F(f) + G(f) \text{ --------------} \qquad (A5.3.3)$$

THE SHIFTING THEOREM

The theorem is: If $F(f)$ is the transform of $f(t)$, then the transform of a function which has the form $f(t-T_1)$, where T_1 is a constant, will have the form $e^{-i2\pi fT_1}F(f)$. We note that, for the functions used in this book, $f(t) = 0$ for $t < 0$ because we have taken the lower limit of the domain to be $t = 0$. This remains true even when the function is shifted, therefore, we may set the lower limit of integration to 0. Substituting (t-T1) for t in (A5.3.1):

$$F(f) = \int_0^{\infty} f(t-T1)\, e^{-i2\pi f(t-T1)}\, d(t-T1) \text{ -----------------} \qquad (A5.3.4)$$

Recognizing that the integrand may be written:

$$f(t-T1)\, e^{-i2\pi f(t-T1)} = \frac{f(t-T1)\, e^{-i2\pi ft}}{e^{-i2\pi fT1}}$$

Then, if we multiply both sides of (A5.3.4) by $e^{-i2\pi fT1}$:

$$e^{-i2\pi fT1}F(f) = \int_0^{\infty} f(t-T1)e^{-i2\pi ft}dt \text{ -------------} \qquad (A5.3.5)$$

The right side of (A5.3.5) is, of course, the Fourier transform of the shifted function f(t-T1).

THE MODULATION THEOREM

We did not illustrate this theorem in the text, but it will be helpful in understanding the Stretching theorem. The theorem relates to the practice of multiplying a function f(t) by a sinusoid in a process generally known as Amplitude Modulation. In radio engineering this process is known to produce two "half amplitude sidebands" centered on the "carrier frequency" f_o. The theorem is: If F(f) is the transform of f(t), then the transform of f(t) $\cos(2\pi f_o t)$ is $\frac{1}{2}$ F(f-f_o) + $\frac{1}{2}$ F(f+f_o).

$$F(f) = \int_{-\infty}^{\infty} f(t)\, Cos(2\pi f_o t)\, e^{-i2\pi ft} dt \quad \text{----------------} \quad (A5.3.6)$$

Recognizing that:

$$Cos(2\pi f_o t) = \frac{e^{i2\pi fot} + e^{-i2\pi fot}}{2}$$

$$F(f) = \frac{1}{2} \int_{-\infty}^{\infty} f(t)\, e^{i2\pi fot} e^{-i2\pi ft} dt + \frac{1}{2} \int_{-\infty}^{\infty} f(t)\, e^{-i2\pi fot} e^{-i2\pi ft} dt$$

$$= \frac{1}{2} \int_{-\infty}^{\infty} f(t)\, e^{-i2\pi(f-fo)t}\, dt + \frac{1}{2} \int_{-\infty}^{\infty} f(t)\, e^{-i2\pi(f+fo)t}\, dt \quad \text{--} \quad (A5.3.7)$$

$$= \frac{1}{2} F(f-f_o) + \frac{1}{2} F(f+f_o) \quad \text{----------------------------} \quad (A5.3.8)$$

Note: $f_o > f$ for normal operation yielding negative frequencies for (f-f_o) implying the "lower sideband" is the complex conjugate of the upper.

THE STRETCHING THEOREM

When we "stretch" a digitized function by placing zeros between the data points, we are actually performing two operations: We are expanding the function and we are introducing modulation. The signal being modulated (i.e. the "carrier") may be represented as a Cos^2 function which produces ones and zeros at the Nyquest frequency f_N. Since a frequency doubling occurs when we square a cosine wave we must use

$f_N/2$ in our equation i.e. $\text{Cos}^2(2\pi f_N/2)$. Our data samples then occur when $\text{Cos}^2(2\pi f_N/2)$ equals 1 and 0.

The theorem states: when dealing with the DFT, if $F(f)$ is the transform of $f(t)$, then the transform of $f(t)_{\text{stretch}}$ will produce the original $F(f)$, plus a duplicate copy of $F(f)$, both of which will be half the amplitude of the original and half the spectrum width.

The transform of the stretched function is represented by:

$$\int_0^\infty f(t/2)\ \text{Cos}^2(2\pi(f_N/2)t)\ e^{-i2\pi ft}dt \quad\text{-----------------}\quad (A5.3.9)$$

Recognizing that:

$$\text{Cos}^2(2\pi(f_N/2)t) = \frac{1 + \text{Cos}(2\pi f_N t)}{2}$$

(A5.3.9) becomes:

$$\tfrac{1}{2}\int_0^\infty f(t/2)\ e^{-i2\pi ft}dt\ +\ \tfrac{1}{2}\int_0^\infty f(t/2)\ \text{Cos}(2\pi f_N t)e^{-i2\pi ft}dt$$

The first part, which we will call $F_1(f)$, yields $F(2f)$ from Similarity. This is the compacted spectrum of the original function. The second part, $F_2(f)$ must be modified to put things in terms of $t/2$:

Recognizing that $\text{Cos}(2\pi f_N t) = \dfrac{e^{i2\pi fnt} + e^{-i2\pi fnt}}{2}$

$$F_2(f)\ =\ \tfrac{1}{2}\int_{-\infty}^\infty f(t/2)\ e^{-i4\pi(f-fn)t/2}d(t/2)\ +\tfrac{1}{2}\int_{-\infty}^\infty f(t/2)\ e^{-i4\pi(f+fn)t/2}\ d(t/2)$$

$$=\ \tfrac{1}{2}\,F(2(f-f_N))\ +\ \tfrac{1}{2}\,F(2(f+f_N))$$

Note that the spectrums of these two functions have both been compressed just as the spectrum of $F_1(f)$ was. To obtain the complete transform we sum $F_1(f)$ and $F_2(f)$:

$$\int_0^\infty f(t)_{stretch}\ e^{-i2\pi ft}dt = \tfrac{1}{2}F(2f) + \tfrac{1}{2}F(2(f-f_N)) + \tfrac{1}{2}F(2(f+f_N))\ \text{---}\quad (A5.3.10)$$

A few words will help greatly to make sense of this. In a non-stretched function, the Nyquest frequency divides the two halves of the spectrum. The spectrum displayed above the Nyquest is comprised of the negative frequencies in a mirror image of the lower positive frequencies. In a stretched function the spectrum is compressed (first term of eqn. A5.3.10) such that the positive frequencies will extend only half way to the Nyquest frequency from the bottom, and the negative frequencies will extend half way from the top of the frequency domain. The last two terms of A5.3.10 provide spectrums centered about the Nyquest frequency similar to the "sidebands" of the Modulation Theorem. The overall result is that the original spectrum is now duplicated.

APPENDIX 6.1

TIME TRIALS PROGRAM LISTING

We don't actually need to write another "core" DFT program, of course; the one from the previous chapter is as good as any other we might come up with (perhaps it is even better for our present purposes since it contains no special provisions to enhance performance). We will need to make a few changes to DFT5.00A to accomplish our objectives however, as listed below:

```
6 REM   ** (DFT6.01A) GENERATE/ANALYZE WAVEFORM **

10 Q = 256

24 PRINT " 1 = TIME TRIALS": PRINT

43 TIM9 = TIMER

45 TIM9 = TIMER - TIM9

47 PRINT "TIME WAS "; TIM9; " SECONDS"
```

In addition to the above changes, we need to write a short routine which will actually perform the time trials (similar to the routines that illustrated the theorems of the previous chapter).

```
300 CLS : REM    *        TIME TRIALS        *
301 Q = 256: REM ******************************
302 FOR I = 0 TO Q - 1: C(1, I) = 0: S(1, I) = 0
304 FOR J = 1 TO 2: KC(J, I) = 0: KS(J, I) = 0: NEXT: NEXT
305 COR$ = "P"
306 GOSUB 900
310 INPUT "ARRAY SIZE AS POWER OF 2"; Q1
312 Q = 2 ^ Q1
324 FOR I = 0 TO Q - 1: C(1, I) = 0: S(1, I) = 0: NEXT
328 C(1, 0) = Q
332 M = 1
336 GOSUB 40: REM TAKE XFORM
340 PRINT "CONTINUE ILLUSTRATION ? (Y/N)"
```

```
342 A$ = INKEY$: IF A$ = "" THEN 342
344 IF A$ <> "N" AND A$ <> "n" THEN 310
346 RETURN
900 CLS : SCREEN 9, 1, 1: COLOR 15, 1: REM TEST DESCRIPTION
902 FOR DACNT = 1 TO 6
904 READ A$: PRINT A$
906 NEXT
908 INPUT "C/R TO CONTINUE"; A$
910 SCREEN 0, 0, 0: RETURN
920 DATA "                TEST 1"
922 DATA " "
930 DATA "In this routine our sole purpose is to illustrate"
932 DATA "the time required to perform a DFT for various lengths"
934 DATA "of input data arrays."
936 DATA " "
938 DATA " "
1000 STOP
```

In the above routine we initialize things in lines 301 through 306. At line 310 we allow selection of the array size (i.e. Q) and then generate an impulse function for our input (lines 324-328). At line 336 we take the transform as usual except that now the time just before and after entering the xform routine are taken (lines 43 and 45 above). The difference between these two times is then printed at line 47 (after printing out the transform results). At line 340 the user is given the option of continuing or terminating the illustration.

APPENDIX 6.2

To incorporate coefficient matrices we must make the following changes:

```
10 Q=64
14 DIM C(2,Q),S(2,Q),K(2,Q,Q)
325 FOR I=0 TO Q:FOR J=0 TO Q
326 ARG=K1*I*J:K(1,I,J)=COS(ARG):K(2,I,J)=SIN(ARG)
327 NEXT J:NEXT I
```

In lines 325 through 327 we generate all of the sine and cosine coefficients we will need in the transform, then, at lines 210 and 211:

```
210 C(M,J)=C(M,J)+C(N,I)*K(1,J,I)+K6*K(2,J,I)*S(N,I)
211 S(M,J)=S(M,J)-K6*C(N,I)*K(2,J,I)+S(N,I)*K(1,J,I)
```

In lines 210 and 211 we change the transform equations to use this array of coefficients rather than computing new sine and cosine values each time through the loop. Also change line 301 to Q=64 and delete line 304.

APPENDIX 6.3

Eliminating the negative frequencies can be accomplished very simply by changing one line in the transform algorithm:

```
206 FOR J=0 TO Q/2:REM SOLVE EQNS FOR EACH FREQUENCY
```

Note that this still includes the Nyquest frequency.

While we are at it we might as well take the time to clean up the "twiddle factor" generation routine at lines 325 through 327:

```
325 FOR I=0 TO Q/2:FOR J=0 TO Q
```

and we can also cut down the size of the K(N,I,J) data array in the dimension statement at line 14:

```
14 DIM C(2,Q),S(2,Q),K(2,Q/2,Q)
```

APPENDIX 7.1

VECTOR ROTATION

A vector may be represented as either a magnitude at some angle or as rectangular components as shown below in figure A7.1. These representations are equivalent—one can easily be converted into the other.

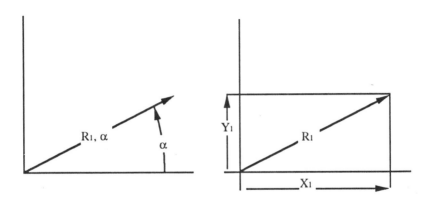

Fig. A7.1 - Polar and Rectangular Vector Representations

Now, it is apparent that, in the polar format (R_1, α), we may rotate this vector through an angle β by simply adding β to α; however, when working in rectangular coordinates, it is not immediately apparent how to accomplish this rotation. We could convert to the polar format, add the angle of rotation, and then convert back to rectangular coordinates but that would be tedious and time consuming.

Actually, it is not difficult to handle rotations in the rectangular format. Consider the rectangular representation of R_1 (as shown in Fig. A7.1 above); we may rotate this whole assemblage of coordinates by the angle β (see Fig. A7.2). The two X,Y components will still add up to the original vector R_1, but now each of these coordinates is a vector itself (X_1, Y_1 in Fig. A7.2). To solve the problem we have set for ourselves we must find the components of each of these two vectors along the X and Y axes.

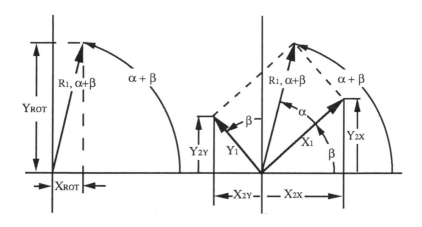

Fig. A7.2 - Rotated Vector Components

Obviously they have been rotated by an angle β, so each of their new X,Y coordinates may be found as:

$$Y_{2y} = Y_1 \cos \beta$$
$$X_{2y} = -Y_1 \sin \beta$$

and,

$$Y_{2x} = X_1 \sin \beta$$
$$X_{2x} = X_1 \cos \beta$$

Now that we have all of these components aligned along the X and Y axes we may simply sum them together to find the components of the rotated vector:

$$Y_{ROT} = Y_{2y} + Y_{2x} = Y_1 \cos \beta + X_1 \sin \beta$$

and,

$$X_{ROT} = X_{2x} - X_{2y} = X_1 \cos \beta - Y_1 \sin \beta$$

which is the relationship we wanted and the one used in Chapter 7.

BIBLIOGRAPHY

If you are not well founded in the calculus there is precious little available in the literature. If you *have* a working familiarity with calculus, and want to continue your study, I recommend the following:

Bracewell, R. N., *The Fourier Transform and Its Applications,* McGraw-Hill. This is my personal preference and my most often used reference. If you become serious about this subject this book will eventually find its way into your personal library.

Brigham, E. O., *The Fast Fourier Transform And Its Applications,* Prentice Hall. This is another standard you will want to become familiar with— extensive bibliography.

Walker, J. S., *Fast Fourier Transform,* CRC Press. Some excellent material—comes with a disk.

There are many other books available and new ones being published every year, but you should judge the value of these for yourself.

Aside from the above, there are a couple of articles you will probably want to read or collect:

Special issues of the IEEE *Transactions on Audio and Electroacoustics* on the FFT, Vol. AU-15, No. 2, June 1967 and Vol. AU-17, No.2, June 1969.

Bergland, G. D., "A Guided Tour Of The Fast Fourier Transform," *IEEE Spectrum,* Vol. 6, July, 1969.

Cooley, J. W., and J. W. Tukey, "An Algorithm For The Machine Calculation Of Complex Fourier Series," *Mathematics of Computation,* Vol. 19, April, 1965. This, of course, was the article that "lit the torch!"

INDEX